REVIVING PALMYRA IN MULTIPLE DIMENSIONS

Published by
**Whittles Publishing Ltd.,**
Dunbeath,
Caithness, KW6 6EG,
Scotland, UK

www.whittlespublishing.com

© 2018 Minna Silver (text copyright ch. 1–10), Gabriele Fangi & Ahmet Denker

ISBN 978-1-84995-296-5

Printed by Melita Press, Malta

# REVIVING PALMYRA
## IN MULTIPLE DIMENSIONS
Images, Ruins and Cultural Memory

Minna Silver
Adjunct Professor in Near Eastern Archaeology,
University of Oulu, Finland

Gabriele Fangi
Professor, Università Politecnica delle Marche, Italy

and

Ahmet Denker
Head, Department of Electrical & Electronics Engineering,
Istanbul Bilgi University, Turkey

Whittles
Publishing

This book is dedicated to the people of Palmyra.

# CONTENTS

# ACKNOWLEDGEMENTS

We wish to thank the Museum of Palmyra, its Director Waleed Al-As'ad, and former Director Generals of Antiquities and Museums of Syria Sultan Muhesen, Tammam Fakouch, A. Moaz and Bassam Jamous for their kind co-operation in the archaeological research carried out in the Palmyra region. Dr. Michel Al-Maqdissi and Prof. Ammar Abdul-Rahman were responsible for the foreign missions there and deserve our gratitude. Dr. Kenneth Silver (former Lönnqvist) has taken a number photographs in this book and created the index for which we are very grateful. Also Ms. Silvana Fangi deserves our thanks for some photographs and for assisting with the index.

*A view towards the Triumphal Arch and the Grand Colonnade of Palmyra photographed in 1920–1930.* Source: Library of Congress, American Colony collection.

# PREFACE

This book provides the means visually to preserve memory and to understand the heritage of ancient Palmyra, a city situated in the middle of Syria, in a green oasis surrounded by a sandy desert. The city has been the focus of international news in recent years and has gone through changing situations in the battles of the Syrian civil war that started in 2011. Both the people and antiquities of Palmyra have faced peril. This book is dedicated to the people of Palmyra and the healing of the place; it is about the heritage of Palmyra that we all share with the Syrians.

The magnificent ruins of Palmyra have become part of our common visual memory through the news and earlier accounts. The place has an extensive history but the site became a battlefield and its inhabitants suffered irreplaceable human losses with many survivors displaced. The conquests and occupations by ISIS/ISIL were brutal interludes that did not respect either people or ruins. Culture was erased and hundreds of years of memory were pulverized by exploding ancient ruined and restored buildings, hammering statues, breaking small objects and destroying mummies.

Recording and documentation form the backbone for saving the memory of a place. Modern technology and imaging methods have provided the means to see how we can approach the city of Palmyra from air and space. Old maps, drawings, paintings, photographs and digital images all contribute to recording and documenting the heritage visually.

Archaeology is a noble discipline that provides deeper and primary knowledge of the past: it reveals to us ruins, objects and new texts. It provides new information about the past. For over a decade the principal author of this book has worked in the area of Palmyra and Deir ez-Zor with the local antiquities authorities. This work precedes the civil war and therefore some of the information provided in this book is first-hand knowledge not available elsewhere.

Personal experience and the studies of the ruins and present life in Palmyra and its surrounding desert-steppe up to the Euphrates provide knowledge that is irreplaceable, that cannot be replaced by second-hand sources. By traversing the same places and landscapes, seeing and sharing the views that the ancients experienced one can experience the past.

Intermediating experience and knowledge is a privilege and in facilitating this process the years of friendship and cooperation with Muhammed Turki and his family from Palmyra have been of special value. In late 2017 Muhammed sent a picture of the reconstructed statue of the lion of Allat that had stood in front of the Palmyra Museum and which ISIS had destroyed. It has a symbolic meaning in healing. Visual reconstruction by means of images and 3D modelling of the ruins of the city are also part of healing process that this book provides for us all.

1

# INTRODUCING THE MEMORY OF THE PLACE

## 1.1. An image for memorising

Every place has its past, its history. There is a sense that is activated when we look at an old drawing, painting, or photograph of a place.[1] If a space, landscape, or site is connected to our knowledge, experience, and memory, we may be able to identify the site. We may attach

Fig. 1.1 *James Dawkins and Robert Wood Discovering the Ruins of Palmyra,* Gavin Hamilton: 1758, National Gallery of Scotland, Edinburgh

Fig. 1.2 Queen Zenobia's Last Look Upon Palmyra, Herbert Gustave Schmalz, 1888

emotions to it, especially if we have experiences associated with the actual place.[2] Pierre Nora states that the memory of places concerns sites where cultural memory crystallises and where our collective memory condenses.[3] We wish that time would stop trying to prevent us from remembering those sites. We have the will to remember.[4] Monuments have become focal points of memory, interpreted as time-marks. This book is about the city of Palmyra in Syria and how its ruins from the Greco-Roman period can be revealed (Fig. 1.1) and revived in our memory through visual representations after the disaster caused by the conquest of Palmyra, the UNESCO World Heritage Site, by the Islamic State of Iraq and Syria (ISIS)/the Islamic State of Iraq and the Levant (ISIL) in May 2015.[5]

The site of ancient Palmyra, situated on the Silk Road in an oasis of the Syrian Desert, is immediately recognisable by its golden ruins that are mainly built from limestone. The city flourished in the first centuries AD. It gained its riches from the caravan trade, especially from textiles, like the silk that was brought from China to the Roman Empire. The Triumphal Arch opened a way to the Grand Colonnade, a street lined with columns, leading to a *Tetrapylon,* a four-column monument at the crossroads. Statues once stood on the column brackets and in the wall niches. Inscribed texts commemorated people of the city and their deeds, their lives, and their gods, like the travellers who dared to move in the area in the 17th to 19th centuries and who had to pay taxes to the local pastoralists such as Bedouins.[6] Westerners are just visitors in a foreign land whose past they can admire and the influences which they have experienced and adapted to their own culture. Palmyra has never enjoyed colonialist attitudes;[7] the site has an oriental pride of its own, a fabulous past and an identity that is recognisable in the dignity of people like former Director of the Palmyra Museum Khaled al-As'ad,[8] who protected its ruins and patrimony with his life.

Palmyra and Tadmor are both names for the same city. Even if western people only recognise Palmyra (Palmyre, Palmira), Tadmor is the original oriental and most likely the original Semitic name of the place. The name Tadmur is still used in the Arabic form today. The Greco-Roman name of Palmyra refers to the palms and groves that make the green oasis in the brown desert so attractive. It has also been suggested that Tadmor comes from the Semitic word *Tamar,* meaning a date palm. Date palms are the speciality of the oasis that has derived its life from the waters of the perennial Efqa spring that have been channelled to its groves and orchards.[9] Others have  suggested different sources for the name, even Hurrian origins. But Tadmor has been and is a Semitic city full of oriental flavour. Its inhabitants have a common descent from the Amorites, Arameans, Arabs, and even Jews, all Semites in their origin. Pastoralism with goats and sheep is in their heritage; caravaneering and trade with donkeys, mules, camels, and horses are embedded in their way of life.

Palmyra was a Roman city that strove for its independence during the reign of the famous Queen Zenobia in the 3rd century AD, even briefly achieving its own empire that reached as far as Egypt and Anatolia. Zenobia identified her roots from Arabs and from Queen Cleopatra herself. She was an oriental ruler. Zenobia's empire minted coins depicting the Queen wearing a helmet. Palmyra has been called 'the Queen of the Desert,' 'the Bride of the Desert,' 'Venice of the Desert,' or 'the Venice of the Sands.' Identifying Palmyra as the 'Queen' or 'Bride' reminds one of Queen Zenobia (Fig. 1.2) as the personification of the city. Indeed, there is something feminine and majestic in Palmyra. In the ancient world cities were often represented by female personifications, such as Antioch (on the Orontes in Syria) identifying itself with the goddess Tykhe. The 'Venice of the Desert' or 'the Sands' echoes the splendour of the architecture that had an enduring impact in the Classical revival, and the sands that surround the site like endless waters. But, the Romans conquered Palmyra, Zenobia, and her empire, and she was taken to Rome in her golden chains. Other conquests and destruction were to follow. (See more in Ch. 5).

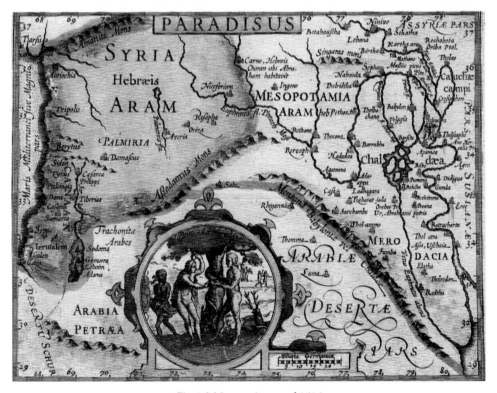

Fig. 1.3 Mercator's map of 1616

Fig. 1.4 *View at the Ruins of Palmyra*, G. Hofstede van Essen, 1691, courtesy: University of Amsterdam

Fig. 1.5 *View at the Ruins of Palmyra*, G. Hofstede van Essen (drawn by Cornelis de Bruijn), 1693

Fig. 1.6 *The Ruins of Palmyra, otherwise Tedmor, in the Desart*, Robert Wood, 1753

## 1.2 Rediscovering Palmyra

Following the Christian bishops and Arab conquerors of Palmyra, a Spanish rabbi called Benjamin Tudela (1160–1173) visited Tadmor[10] at the time of the Crusades. He reports that he first visited Baalbek then Tadmor: 'this city is surrounded by a wall, and stands in the desert, far from any inhabited place, being four days' journey distant from the above-mentioned Baalath [Baalbeck].' From the account of his itinerary hundreds of years passed in sleep before more western travellers entered the city. The word 'renaissance' means the rebirth or the revival of the Classical past in the 15th and 16th centuries; it was visible in literature, art, fashion, and in the philosophy of the whole era. It was a time of the revival of the ancient worlds, the ruins, and antiquities. It was also a time of great explorations that widened our world view. New continents and sites were discovered. Although Palmyra had not yet been revisited, Mercator's map from the year 1616 (Fig. 1.3) clearly shows that there was knowledge of its location, apparently based on the itineraries and maps from antiquity, such as Ptolemy's Geography from the second century AD. In Mercator's map, however, Palmyra is associated with the Biblical paradise in Mesopotamia, the land of the Euphrates and the Tigris.

It took some time to 'rediscover' Palmyra. Thirty men arrived in Palmyra in 1691,[11] reviving the study of Palmyra and its remains. A large panorama of the ruined city was produced as a painting in Holland in 1691 and a drawing of the same view was produced in 1693 (Figs. 1.4 and 1.5). When Palmyra was 'rediscovered', Albinoni composed an opera named *Zenobia Queen of the Palmyreans*.[12] Englishman Robert Wood (Fig. 1.1) published in 1753 *The Ruins of Palmyra, otherwise Tedmor, in the Desart* (London) reproducing the city and its ruins in drawings, including panoramas. The drawings show decorative details of architectural elements from Palmyra that had fallen into ruins. Explanations of the views were also provided.[13] (Fig. 1.6) Consequently, Louis François Cassas produced beautiful drawings and paintings of the

Fig. 1.7 *A miniature model of the ruins of Palmyra, Betty Ratcliffe 1765–1780,* National Trust

buildings and landscapes of the city in *Voyage pittoresque de la Syrie, de la Phénicie, de la Palestine et de la Basse Egypte* in 1798–1804 (Paris, D'Imprimie de la République).

Palmyra was there, already partly visible on the ground surface, to be rediscovered and documented by the Europeans. But for eastern people, such as Arabs, the city was still alive in the Temple of Bel. The same applied to Petra, another caravan city of the Nabatean kingdom, which Johann Ludwig Burckhardt, a Swiss explorer, rediscovered in 1812.[14] The site, with its facades of rock-cut tombs, opens up through the Siq entrance, veiled between high reddish natural walls of sandstone cliffs. These sites had always been there to be rediscovered by the west. Instead, Pompeii and Herculaneum in Italy became sites for archaeological excavations in the 18th century. They were places that had to be exposed from the earth. Archaeology was born when the removal of layers of ash and pumice revealed complete houses, streets, and life that had stopped in the eruption of Vesuvius in AD 79. [15]

Classical archaeology came about through Johann Joachim Winckelmann and his followers in publications that included Pompeii and Herculaneum in the 1760s. *Geschichte der Kunst des Altertums* (The History of the Art of Antiquity) became Winckelmann's *magnum opus*, a great work of art. After the French Revolution Napoleon, with his engineers, added descriptions of Egyptian ruins to European knowledge; the taste of the new 'empire' style continued into the 19th century. Architecture in western cities derived its styles from Greece and Rome, and also from the ruins of Palmyra. Palladian and Neo-Palladian villas followed the ideas of the Pantheon with its dome in Rome, and the Capitol of Washington carries visible influences from Palmyra. Rich floral elements in the decorative art of *stucco* were copied.[16] Everybody could have a cultural memory of Palmyra in their surrounds. Artist Betty Ratcliffe built a delightful miniature model of Palmyra and its ruins in 1765–1780, shortly after Robert Wood had published his book. The model was made from mother of pearl, stone, and glass and was kept in a glass case; now it is to be found in the collection of the National Trust at Erddig, Wrexham, North Wales (Fig. 1.7). It seems to represent the Temple of Bel area following the engravings given in Wood's book. Palmyra became every man's dream – and the South Wales city of Swansea even has a road called 'Palmyra Court'.

Daring female traveller Lady Hester Stanhope (1776–1839) travelled from Britain to the Near East in 1810, when Napoleon's interest in the Orient was at its peak. She proceeded to Palmyra with her caravan.[17] She broke the general rules set for ladies, as she travelled without official chaperones and wore Oriental clothing whilst riding on horseback. She has also been called 'Queen of the Desert' or 'Queen of the East'.[18] By the end of the century, travels in the east to explore the ruins of Palmyra similarly inspired Gertrude Bell, another 'Queen of the Desert,' an archaeological explorer and traveller from Britain. Bell visited Palmyra in 1900, taking black and white photographs of the ruins. She exclaimed in her letters: 'Palmyra [Tadmur], for I've got here at last, though after such a ride!' (20th May, 1900). Photography had started a new era and way of documentation in the 19th century; photographs were also taken from the air from balloons and kites, then later from aeroplanes. Palmyra was photographed from the air by the French during the Mandate era, and the collections of views of the ruins were created.

## 1.3. The ruins, Palmyra Museum, and the guest house

The archaeological landscape of ancient Palmyra is especially famous for its tower tombs (Fig. 1.8). Upright walls of buildings are left standing on the plains. Numerous columns, pilasters, and decorated capitals with Ionian volutes or those in the Corinthian style, full of acanthus leaves or mixed with Egyptian style flowers are found in the ruins. There are also fallen column drums and decorated architraves, and Assyrian merlons (stone triangles) reminiscent of small ziggurats (Mesopotamian stepped temple forms) scattered on the ground.

Palmyra has its own language of style that mixes Greco-Roman orders with an Oriental flavour from Syria, Mesopotamia, Persia, and Phoenicia. This mixture is very Hellenistic in its essence – Alexander the Great wished to impose on the Near East by fusing Greek and Oriental cultures and worlds. The Aramaic language came to be used in the Persian world as a *lingua franca*. It was followed by the Greek language as another *lingua franca* or *koiné* in the Hellenistic world, which continued until the Roman occupation. Both languages were still used in Roman Palmyra, as in the world of Jesus and his Apostles in Roman Judea, alongside the official Latin. Today, Aramaic is still spoken in its Syriac form in Syria.

The National Museum of Palmyra, situated in the modern part of the city, was opened in the 1960s. There was a miniature of the reconstructed Temple of Bel (Fig. 1.9), the most imposing building in Palmyra, on display. Reliefs from tombs hung on the walls, and large statues stood in the halls. Glass cases were filled with artefacts dating from the Palaeolithic Age (the Old Stone Age) onwards.

SYRIA, PALMYRA, VALLEY OF THE TOMBS, OKT. 2006

Fig. 1.8 *Dutch artist José Lips Besselink painted the famous Valley of the Tombs in 2006 and lived in the Guest House of the archaeologists in the precinct of the Temple of Bel.*

Fig. 1.9 *A miniature model of the Temple of Bel in the Museum of Palmyra*
Photo: Kenneth Lönnqvist, 2004

Fig. 1.10 *A postcard from a Palmyrene lady Sumaya Turki to Minna Silver. Sumaya Turki is married to Muhammed Turki, a cousin of Waleed al-As'ad. The Turki family had to flee ISIS and their house was destroyed.*

Fig. 1.11 *The guest house for archaeologists working in the Palmyra region inside the precincts of the Temple of Bel.* Photo: Gabriele Fangi 2010

There was plenty of pottery, Roman glass, and jewellery. The upper floors and the basement were used for storage, and only authorised persons could enter those areas. Inscriptions were displayed in the garden. On the ground floor there were offices of the director and his representatives. The Al-As'ad family – Khaled Al-As'ad and his son Waleed – had been in charge for the museum for decades. The principal author of this book worked under both directors and spent time in their home. Khaled Al-As'ad's son Muhammed and his cousin Muhammed Turki worked as local representatives for the author's project (Fig. 1.10).

Many archaeologists have stayed in the guest house of the Directorate-General of Antiquities and Museums (DGAM) of Syria in the Temple of Bel precincts over the decades, including the expedition of the present principal author. It is a wonderful traditional Arab house, with large thick doors, arches, and a central courtyard with palm trees, reminiscent of *A Thousand and One Nights* (Fig. 1.11). On the upper floor you could examine your small finds in study rooms or sleep in private rooms. In the evening you could sit, or if hot even sleep, on its roof, gazing at the ruins of the temple building itself. It was said that the Emperor of Japan had admired the ruins and the sunset from the guest house roof (Fig. 1.12).

## 1.4. Destroying memory

Louis François Cassas documented tomb raiders in action in Palmyra in his engraved drawings of the tower tombs in the Valley of the Tombs in the 1790s.[19] In Gertrude Bell's photographs there are images that show how much looting of the ruins had already taken

Fig. 1.12 *The sun setting behind the courtyard of the Temple of Bel in Palmyra.*
Photo: Minna Lönnqvist 2008

Fig. 1.13 *Early restoration work in Palmyra. Workmen are moving a massive capital for a column. Photographed in 1920–1933.* Library of Congress, the American Colony collection

place in the Valley of the Tombs in 1900. Many statues were taken from Palmyra and transported to homes and museums around the world, even in the 20th century. Restoration work also took place in Palmyra (Fig. 1.13). The archaeological and conservation debates of today deal with ethical issues concerning removal of funerary monuments from their original sites. In the face of war and pillage the transportation of items to a safer place might be a necessity as in the case of Palmyra on the eve of the invasion in 2015.

The conquest of Palmyra by ISIS in May 2015 led to fears that many people hoped would not be realised, after what had already happened at Hatra, Nineveh, and Nimrud in Iraq. It was anticipated that the ruins of Palmyra would not face that fate of destruction of buildings and statues that had been witnessed in Iraq but would have a special, more favourable place in the minds of the jihadists.[20] The peril was, however, looming.[21] After the conquest of the city, atrocities and destruction began in the city's ruins, and the Museum's artefacts were damaged and looted. Numerous, illicit diggings were traced by satellite imagery.

The outer walls of the Temple of Bel had already been damaged by air strikes and shelling during the civil war that broke out in 2011, before ISIS entered the city. Destruction and pillage in the offices inside the Temple of Bel courtyard also took place during the first years of the civil war; soldiers looted antiquities, and along with ISIS the destruction became part of our collective memory. The Temple of Baalshamin, the Temple of Bel, the Triumphal Arch, and tower tombs were blown up in 2015. The Museum's artefacts were damaged or destroyed.[22] In 2017 the *Tetrapylon* and part of the theatre were destroyed. Images displaying the destruction of the monuments, such the Temple of Bel, were televised via satellites and streamed on the internet. People were executed in the ancient theatre as a spectacle. Both directors of the Museum, the retired Director Khaled Al-As'ad and current Director Waleed Al-As'ad, became victims of ISIS: Khaled was brutally executed in August 2015,[23] and Waleed was imprisoned and tortured.[24] Both had been interrogated about the hidden treasures of Palmyra.

Destroying monuments meant the destruction of cultural memory. Iconoclasm is not new but it was rediscovered by ISIS and reinterpreted. It means demolishing images of humans or gods that according to religious rules should not be portrayed. By devastating ruins, defacing or decapitating statues and humans, ISIS tried to deface our memory and rule our minds. Artefacts from the Museum were also looted, just as the Syrian troops had done earlier. Iconoclasm was reborn and had taken over. That was not, however, a new religious or political approach invented by ISIS. Mesopotamia had gone through such phases in the ancient past when statues were decapitated and defaced for religious and political reasons.[25] There is also evidence that such destruction had taken place in Palmyra after the Roman period as several statues appear to have been beheaded on purpose (Fig. 1.14).

Biblical accounts tell of destroying idolatrous statues and images. Much later on in the Byzantine Empire, an iconoclastic controversy broke out between those supporting icons and those who wanted to destroy them. At the beginning of Islam, icons presenting humans or human-shaped gods were criticized by Muslims for religious reasons, especially in the 8th century AD. The idea of 'no graven images' had also applied earlier among the Jews.[26] Later on Protestants covered church frescos of the Catholic era with lime plaster – new Protestant creeds did not approve worship of statues or icons in the 16th century. The year 1566 was the

Fig. 1.14 *Decapitated statues and sarcophagi, photographed in Palmyra in front of the Valley of Tombs in 1920–1933.* The Library of Congress, the American Colony collection

peak of iconoclasm in the reformation movement known as Beldeerstorm in Holland; similar activities took place not only in the Continent but also in Britain.[27] The Taleban following the Islamist ideas blew up the Bamiyan Buddhas at a World Heritage Site in Afghanistan in 2001.[28]

*Damnatio memoriae*, the cursing and erasing of the memory of a certain person, has also been used since ancient times. There may be a belief that such actions affect the person in a magical way. In the case of dead people they were made 'completely dead'. Portraits and inscriptions have been destroyed for political reasons. Joseph Stalin used to erase disfavoured people from photographs. In the course of the collapse of the Soviet Union in 1991 statues of Stalin and Vladimir Lenin were destroyed or removed. In the aftermath of the western invasion of Iraq in 2003, the statue of Saddam Hussein in *Firdos Square*, Baghdad was pulled down and attacked with a sledgehammer.[29]

## 1.5. Digitally reconstructing the ruins and objects

Now modern technology, with digital photographs, photogrammetry, and 3D modelling, can provide some substitute for the memories of the lost or destroyed things. Images provide a way to produce remains in 3D (Figs 1.15 and 1.16). Monuments and artefacts which represent common heritage can be brought to life to reawaken memories of what has been lost. The new visual recovery of Palmyra could be seen as its second rediscovery after the 17th century. Modern technology can never replace or totally restore the actual places,

Fig. 1.15 *The ruins of Palmyra in 1932.* Source: A. Poidebard, 1934

Fig. 1.16 *The virtual image of the Temple of Bel in Palmyra. This virtual reconstruction suggests how the temple might have looked in antiquity.* Model: Ahmet Denker 2016

monuments, and artefacts that were destroyed or stolen, but it can revive our memory and help us to experience how the things may have looked. We can recreate the documented monuments, add texture, and augment the monuments with the real surroundings as well.

Therefore, 3D models and reconstructions can provide some new aspects and experiences of a past place and space. They can bring some delight that can gradually provide positive hope and feelings or healing we can all share. *The New Palmyra*,[30] *The Virtual Palmyra*[31] (Fig. 1.16), *Syrian Heritage*,[32] *Rekrei* under Project Mosul,[33] and *Arc/K Project*[34] are all initiatives that have taken the digital remodelling of the ruins or the past Palmyra as their goal. The *New Palmyra* is an open-source project in which audiences can participate via the internet in recreating Palmyra. This virtual Palmyra has its sources in ancient texts, lithographs, old documentation, and archaeological reports. Inspiration has also come from the texts of ancient authors. *Revive Palmyra* is a joint endeavour of ICONEM in France and the Directorate-General of Antiquities and Museums in Damascus, Syria. One can visit the site of the Temple of Bel and see the destruction after ISIS and study pieces online.[35]

3D printing technology has been also used in recreating monuments of Palmyra. The Institute for Digital Archaeology based in Oxford has supplied Syrians with 3D cameras to document monuments, producing millions of images. Crowdsourcing is a new way to model monuments and objects in 3D. The Institute produces models of destroyed monuments from images, for example, the Monumental or Triumphal Arch, thus providing us with some concrete feeling of the monument in its solid shape.[36] It has become a commemorative monument that tours around places associated with suffering. First it was erected in the Trafalgar Square in London followed by a display in New York in 2016.[37] It is a statement to underline that what the jihadists have destroyed we shall rebuild. A full-size replica can provide some dimensional qualities and ideas of an object but can never replace and embody the life-cycle of the original monument or object that carries authentic memory.

In any event, destruction is part of our memory – moments in history. However, as said, we can revive the monuments and cultural objects in our memory using visual aids, from old drawings, paintings, and photographs to digital images, laser scans, and 3D modelling. Digital images and millions of points in point clouds produced by laser scans can be used to recreate the monuments and their past virtually. In 2016 ICOMOS (International Council for Monuments and Sites) under UNESCO arranged meetings focusing on post-devastation reconstructions. Old drawings, images, metric documentation, and 3D models can also help the restoration work on the ground.

## Endnotes

1   See, for example, Trigg, D. (2012) *The Memory of Place: A Phenomenology of the Uncanny.* Series of Continental Thought. Athens: Ohio University Press.

2   See Trigg, D. (2012) *The Memory of Place: A Phenomenology of the Uncanny.* Series of Continental Thought. Athens: Ohio University Press; see also Silver, M. (2016) A Digital Image as a Kaleidoscope, in *Ready to Reach Out: Connecting Cultural Heritage Collections and Serving Wider Audiences,* EU 2016, see: https://issuu.com/cre-aid/docs/minocw-cultural-heritage, pp. 58–59. Accessed 26th September, 2016.

3   Nora, P. (1989) Between Memory and History: Les Lieux de Mémoire, in *Representations,* No. 26, Special Issue: Memory and Counter-memory (Spring), pp. 7–24, especially pp. 7–9.

4   Nora, P. (1989) Between Memory and History: Les Lieux de Mémoire, in *Representations,* No. 26, Special Issue: Memory and Counter-memory (Spring), pp. 7–24; see Van Dyke, R.M. and Alcock, S.E. (ed.) (2003) *Archaeologies of Memory.* Oxford: Blackwell; see also latest discussions in Morcillo, M.G., Richardson, J.H., Santangelo, F. (ed.) (2016) *Ruin or Renewal? Places and Transformation of Memory in the City of Rome.* Roma: Edizioni Quasar.

5   *The Guardian* 21st May, 2015.

6   See, for example, Browning, I. (1979) *Palmyra.* London: Chatto & Windus.

7   See Said, E. (1978) *Orientalism.* New York: Pantheon books.

8   In 2003 the present principal author (Minna Silver, formerly Lönnqvist) sat in Director Khaled Al-As'ad's office when a foreign western woman walked in and told him how he should run the office and take care of the ruins in Palmyra.

9    Hammad, M. (2010) *Palmyre: Transformations urbaines, Développement d'une ville antique de la marge aride syrienne*, Paris: Geuthner, pp. 10–11.

10   Tudela, B. (Repr. 1907) *The Itinerary of Rabbi Benjamin Tudela: Travels in the Middle Ages*. New York: Philip Feldman.

11   Astengo, G. (2016) *The Rediscovery of Palmyra and its Dissemination in Philosophical Transactions*. Notes and records. Royal Society Publishing. Published on line; Browning, I. (1979) *Palmyra*. London: Chatto & Windus, p. 53.

12   Bäärnhielm, G. (1988) Palmyra – en bakgrundsteckning, in *Palmyra, Öknens drottning*, ed. by Hellsröm, P., Nockert, M. and Thålin-Bergman, L., Stockholm: Medelhavsmuseet och Statens historiska museum, pp. 11–30.

13   Wood, R. (1753) *The Ruins of Palmyra, Otherwise Tedmor in the Desart*. London: [Wood].

14   See Burckhardt, J.L. (1822) *Travels in Syria and the Holy Land*, London: John Murray, pp. 422–433.

15   See, for example, Harris, J. (2007) *Pompeii Awakened: A Story of Rediscovery*. London, New York: I.B. Tauris & Co. Ltd.

16   See Browning, I. (1979) *Palmyra*. London: Chatto & Windus.

17   Meryon, C. (Repr. 2012) *Travels of Lady Hester Stanhope, Forming the completion of her memoirs*. Vols. I–III. Cambridge: Cambridge University Press.

18   See, for example, Gibb, L. (2005) *Lady Hester, Queen of the East*. London: Faber & Faber Ltd.

19   See Browning, I. (1979) *Palmyra*, London: Chatto & Windus, p. 74.

20   https://www.theguardian.com/world/2015/apr/05/isis-video-confirms-destruction-at-unesco-world-heritage-site-on-hatra; https://www.theguardian.com/world/2015/mar/09/iraq-condemns-isis-destruction-ancient-sites. Accessed 20th September, 2016.

21   Silver (Lönnqvist), M. (2015) Palmyra – the Peril of the Ruins is Looming, in *CIPA Newsletter* Issue 06, May; Silver, M. (2015) The Destruction of Cultural Memory in Palmyra, in *CIPA Newsletter*, Issue 07, October.

22   See the UNESCO assessment of the damage 27th April, 2016 in http://www.whc.unesco.org/en/lis/23 under "Palmyra". Accessed 25th September, 2016.

23   https://www.theguardian.com/world/2015/aug/18/isis-beheads-archaeologist-syria. Accessed 20th September, 2016.

24   Personal communication with Waleed al-As'ad's cousin Muhammed Turki, a resident from Palmyra.

25   May, N. (2010) Decapitation of Statues and Mutilation of the Image's Facial Features, in *A Woman of Valor: Jerusalem Near Eastern Studies in Honor of Joan Goodnick Westenholtz*, ed. by Horowitz, W., Gabbay, U. and Vukosavoviç, F., Biblioteca del Próximo Oriente Antiquo 8, Spain: CSIC, pp. 105–118.

26   See, e.g., Moorhead, J. (2001) *The Roman Empire Divided, 400–700*. Harrow, London: Longman, Pearson Education, pp. 233–234.

27   Phillips, J. (1973) *Reformation of Images: Destruction of Art in England, 1535–1660*. Berkeley: University of California Press.

28   *Reuters*, 2 March, 2001.

29   The act was televised and is presented in the YouTube. The YouTube accessed 26th September, 2016.

30   See, for example, Busta, H. (2015) An Open-Source Project to Rebuild Palmyra, in *Architect Magazine*, October 23, 2015.

31   Denker, A. (2016) The Virtual Palmyra: 3D Reconstruction of Lost Reality, in *Arqueologica 0.2., 8th International Congress: Advanced 3D documentation, modelling and reconstruction of cultural heritage objects, monuments and sites, 5–7 September, Valencia, Spain*, 2016; see also Denker, A. (2017) Rebuilding Palmyra virtually, in *Virtual Archaeology Review*, Vol. 8, pp. 20–30.

32   https://sketchfab.com/models/02c4e194c6d64a4385a30990ed9899bf. Accessed 26th September, 2016.

33   https://projectmosul.org/. Accessed 26th September, 2016.

34   http://arck-project.com/. Accessed 26th September, 2016.

35   https://sketchfab.com/models/02c4e194c6d64a4385a30990ed9899bf. Accessed 26th August, 2016.

36   Brown, M. (2015) Palmyra's Arch of Triumph recreated in Trafalgar Square, in *The Guardian*, 19th April, 2016.

37   https://www.theguardian.com/us-news/2016/sep/20/palmyra-arch-syria-new-york. Accessed 20th September, 2016.

2

# APPROACHING PALMYRA FROM AIR, SPACE, AND BY LAND

## 2.1 Aerial archaeology and satellite imagery

The aerial capabilities and scope of exploring Palmyra and its surroundings have changed since the first European travellers rediscovered the site and its ruins. Initially there were drawings and paintings, and subsequently photographs appeared from the 19th century onwards. From the beginning of the 20th century some of the first steps in the field of aerial archaeology were taken in the area of Palmyra and the Roman frontier zone of the Eastern Limes in general.[1]

Like L.W.B. Rees[2], Sir Aurel Stein,[3] O.G.S. Crawford, and other pilots of the First World War, Father Antoine Poidebard, a French priest, became a pioneer in aerial archaeology. He documented sites such as Palmyra by photographing them from his aeroplane in Syria during the French Mandate period (Fig. 2.1). He used the shadows of the mornings and the sun in the late afternoon as well as the autumn rain, when the vegetation was the optimum green for tracing sites and monuments. The colour of surface varies due to the effect of buried remains on the condition of the soil above which in turn changes the growth of vegetation in the areas of walls and other ruins.[4]

Visibility and preservation are factors that affect the potential to reveal ancient remains and traces left

Fig. 2.1 *Poidebard's pioneering studies of the Roman eastern border from the air.* Source: A. Poidebard, 1934

by humans.[5] Archaeologists can trace sites and structures by remote sensing with aerial photographs, satellite images, and LiDAR. Remote sensing has opened up the possibility for ways to approach distant sites like Palmyra from the air (e.g. the early aerial photograph of the Palmyra oasis in Fig. 1.15) and space, but actual visits and surveys on the ground are needed to confirm the interpretation and collect finds for dating purposes. An area such as the Syrian Desert and steppes are favourable for archaeological remote sensing because of meagre vegetation cover and optimal weather conditions for most of the year as clouds do not hinder the visibility.

Fig. 2.2 *The oasis of Palmyra photographed in the 1920s/1930s. Library of Congress, Matson Collection*

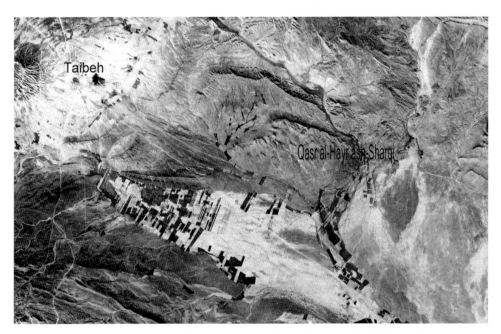

Fig. 2.3 *Taibeh (ancient Oriza) and Qasr al-Hayr ash-Sharqi (ancient Adada?) in the hinterland of Palmyra along the Strata Diocletiana seen from a CORONA satellite photograph from the 1960s.* Remote sensing by Minna Lönnqvist 2008, SYGIS

Google Earth is nowadays the most common way to use satellite imagery for prospecting and mapping. However, the software does not provide properties for mapping, analyses, and classification required for more professional work which requires tessellated images (these are like mosaics carrying huge amounts of information in numeric form). There are valuable satellite data sources that can illustrate the situation of the region of Palmyra in the 1960s, such as declassified CORONA satellite photographs obtained from the CIA's archives. The photographs were taken on black and white film, and when digitised can reach a resolution on the scale of 1.8 metres.[6] Studies of agricultural village life and Bedouin habitation can be carried out from old black and white aerial photographs (Fig. 2.14) as well as from these early satellite photographs covering the region of Palmyra (Fig. 2.3). In the Palmyrene desert, beside Palmyra itself, there are a number of green oases watered by springs and wells previously photographed by Father Poidebard, but now the tessellated multispectral images provide a better way to learn more about the environment.

Multispectral Landsat satellite imagery is especially well suited for environmental studies; it reveals in colour natural features of the Palmyra area, especially the contrasts of the green

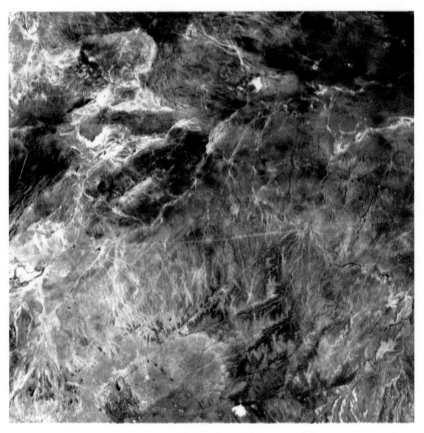

Fig. 2.4 *Landsat-2 MSS-image 185/36 (WRS 1) date 27.06.1975, channel 574. The image reveals caravan roads leading from Palmyra to the Euphrates. Palmyra can be distinguished in green and purple colour on the left edge of the image.* Markus Törmä, SYGIS

Fig. 2.5 *Landsat-7 ETM-image 172/36 (WRS-2), date 26.05.2001, channel 741. Heavy desertification has covered the Syrian Desert and the lines of the caravan roads. The oases of Palmyra and El Kowm can be distinguished on the left from the turquoise colour in the desert.* Markus Törmä, SYGIS

Fig. 2.6 *Landscape model in 3D of the desert-steppe in Jebel Bishri on the Palmyrides constructed by fusing QuickBird image data with ASTER-DEM data.* Constructed by Markus Törmä 2011, SYGIS

oases with the brownish desert. Seasons, such as spring, can be detected through the greenness of the desert, and small agricultural endeavours become visible. Beside desert and steppe environments, the Palmyrides mountains form a defining, majestic feature in the landscape surrounding Palmyra (Figs. 2.4, 2.5, 2.6).

Satellite images can be used in modelling landscapes in 3D by fusing them with radar data and thus showing topographical variations in altitude. Fusing high resolution images such as QuickBird (spatial resolution of 0.6 metres) with radar data can provide information on sites and produce very natural landscape models (Figs. 2.6, 2.7). Like QuickBird (the production of which has ceased), GeoEye is a high resolution multi-spectral satellite imagery system (spatial resolution of 0.46 m) that reveals archaeological sites and structures in their environment.[7]

Classification, like clustering analysis, can reveal in satellite imagery features that are not necessarily visible to the naked eye (Fig. 2.8). Analyses carried out with satellite imagery, radar data, and geophysical studies on the ground have been used in archaeology to see beneath the ground surface. Nowadays LiDAR from an aerial platform provides a new range-based way to see through vegetation and under the ground.[8] Before the civil war broke out in Syria in 2011 the General Organisation of Remote Sensing in Syria (GORS) had produced an archaeological satellite atlas that has been a delightful source for any archaeologist or historian working in the region.

Fig. 2.8 *Clustering analysis made with a Landsat image by extracting grazing potential in the Palmyrides visible in greenness in spring time.* Analysis by Markus Törmä, SYGIS

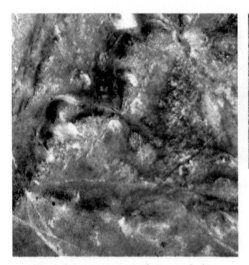

Fig. 2.7 *Ancient pastoral sites including curvilinear corrals, animal pens, and cairn/ tumulus tombs on Jebel Bishri in the Palmyrides, seen from a QuickBird satellite image.* Prospecting by Minna Lönnqvist, SYGIS

## 2.2 Tracing mobile life and sedentarisation

Mobility has been a way to cope with desert and steppe environments. People moving in deserts had to depend on scattered water sources, so often water is the initial sign of human habitation. Springs and wells become essential focal points in movement. First hunter-gatherers, then nomads and caravan traders were active desert and steppe dwellers. A cave, a hut, or a tent (Fig. 2.12) were and still are dwellings of the mobile cultures.

Human presence in the district of Palmyra dates back to the Palaeolithic era, that is to the Old Stone Age – a time of hunting and gathering. The oasis of the El-Kowm basin in the Palmyrene desert (Fig. 2.9) has offered evidence of habitation during the earliest phase, starting from the Lower Palaeolithic pebble tools dating c. 1 million years; these are comparable with the African pebble tool cultures. The Acheulian industry is also present.[9] The desert of Palmyra provides high-quality flint and is full of Middle Palaeolithic tools, representing the Levallois–Mousterian tradition and its local variant. In the surroundings of Palmyra, there is also the famous Douara cave that was inhabited in the Palaeolithic. Its flora and fauna show that in the area that now belongs to desert the climate was moister in the Middle Palaeolithic period.[10]

Gazelle hunting was a common activity in the Palmyra desert during the Stone Age,[11] and also in later periods. Large V-shaped traps for hunting gazelles have been found in numbers in the region;[12] they were first used in the Neolithic (New Stone Age) period. They could later have been used as animal pens by pastoralists.[13] These traps, known as 'kites' (see Fig. 2.10), were first recognised in Jordan by early aerial studies,[14] and their distribution extends from Sinai to Armenia. In this distribution a particular concentration is visible in the Syrian Desert around Palmyra.[15]

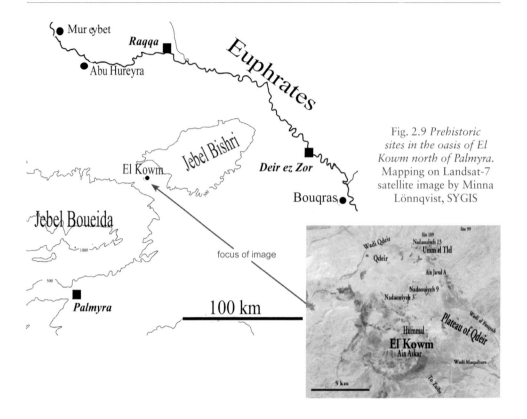

Fig. 2.9 *Prehistoric sites in the oasis of El Kowm north of Palmyra.* Mapping on Landsat-7 satellite image by Minna Lönnqvist, SYGIS

Fig. 2.10 *A kite, a large animal trap, near Damascus photographed by Poidebard from air.* Source: A. Poidebard, 1934

Fig. 2.11 *A crescent-shaped hunting hide from the Palmyrides.* Photo: SYGIS 2000

Johann Ludwig Burckhardt described gazelle hunting and kites in his travels in the Syrian Desert in the early 19th century:

> Gazelles these are seen in considerable numbers all over the Syrian Desert. On the Eastern frontiers of Syria are several places allotted for the hunting of gazelles; these places are called masiade. An open space in the plain of about one mile and a half square, is enclosed from three sides by a wall of loose stones, too high of the gazelles to leap over. In different parts of this wall gaps are purposely left, and near each gap a deep ditch is made on the outside. The enclosed space is situated near some rivulet or spring to which in summer the gazelles resort. When the hunting is to begin many peasants assemble and watch till they see a herd of gazelles advancing from a distance towards the enclosure, into which they drive them; the gazelles frightened by the shouts of these people, and the discharge of fire-arms, endeavour to leap over the wall, but can only effect this at the gaps, where they fall into the ditch outside, and are easily taken, sometimes by hundreds. The chief of the herd always leaps first, the others follow him one by one. The gazelles then taken are immediately killed, and their flesh sold to the Arabs and neighbouring Fellahs.[16]

The skeleton of a wild ass with a Levallois point provides evidence of the hunting of asses in the Middle Palaeolithic period in the area of El Kowm.[17] Wild camels, ostriches, and rabbits belong to the repertoire of the animal life as well. Remains of a giant camel have been found in the area.[18] Crescent-shaped hunters' blinds built of field stones are also numerous and are set along the routes of animals (Fig. 2.11).[19]

Sedentary life started in the Neolithic period c. 11,000–9,000 years ago in the Near East along with agriculture and animal husbandry. Not far from Palmyra on the Middle Euphrates, the first signs of agriculture in the Near East are attested at the village of Abu Hureyra (Fig. 2.9) dating from the Epipalaeolithic and Neolithic periods c. 9,000 BC.[20] The first signs of animal husbandry come from Chawi Shemi Shanidar in Iraqi Kurdistan[21] and at Ali Kosh

Fig. 2.12 *A Bedouin in his tent at the edge of the Palmyrides in the Euphrates valley.*
Photo: Gabriele Fangi 2010

Fig. 2.13 *Bedouin life between a tent and a house.* Photo: Minna Silver 2015

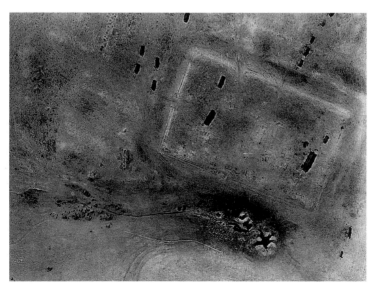

*2.14 The oasis of Qdeir along the Strata Diocletiana between Palmyra and the Euphrates photographed from the air by Antoine Poidebard. A Bedouin camp occupies the site that has provided early evidence of pastoral nomadism dating from the Neolithic period. Source: A. Poidebard, 1934*

in Iran in the Zagros mountains.[22] Agriculture and sedentarisation proceeded hand in hand, and small villages were born. The earliest evidence of pastoral nomadism is to be found in the El Kowm basin in the playas beneath the Palmyrene mountain belt at sites such as the village of Qdeir which dates to c. 7,000–5,000 BC in the Neolithic period.[23] There village-based seasonal pastoral transhumance between the lowlands and highlands follows the pattern of the earlier hunter-gatherers' movements for looking for prey in the mountainous regions. However, even if transhumance is typical for the region, it is not the only type of pastoral nomadism in the area. There are various degrees of nomadism to be found in the past and present Palmyrides and the Syrian Desert.[24] The continuity of pastoral nomadic life is visible in aerial photographs taken by Poidebard (see Figs. 2.12–2.14).

Earlier it was thought that nomads do not leave archaeological traces; they were thought to have been invisible. The ability to detect pastoral life in archaeological record has, however, changed. New methods were developed in the 1980s and especially in the 1990s based on ethnoarchaeological observation. Those observations provided means to detect and make invisible nomads visible. Numerous animal pens and cairns/*tumuli* cover the desert areas and the Palmyrides (Fig. 2.7), demonstrating the presence of pastoral nomadism and showing its impact as a major subsistence economy of the region throughout millennia. Such structures of pastoralists can be detected by high resolution satellite imagery. The cairns/*tumuli* have formed a common grave structure in the area for thousands of years, especially for the pastoral nomads. Recent studies in the desert surrounding Palmyra have shown that although some date from the Late Neolithic period;[25] they became common later in the Chalcolithic and Bronze Ages. Present-day Bedouins in the region still use them. The structures and spatial arrangement of such cairn/*tumulus* fields seem to represent tribal and family structure. Certain features, for instance the ring-*tumulus*-type, may indicate a chiefdom system. Often cairns/*tumuli* are located on hills or mountain edges in the region and have been used as nodes and signposts in the landscape for caravans.[26]

Agriculture and hunting are subsidiary economies for pastoralists in the region, but also collecting of desert truffles (*Terfeziaceae*) and producing salt in the salt pans of the desert have provided livelihoods and possibilities for trading.

## 2.3 Caravan roads in deserts, steppes, and coastal regions

Caravaneering is mobile life and often appears as an offshoot of pastoral life. Tadmor (Palmyra) was already a caravan city during the Bronze Age (3,300–1,200 BC). The caravan roads were the early ways to approach the town over the desert. Therefore, the caravan trade between Mesopotamia and Tadmor did not emerge as a new phenomenon in Greco-Roman antiquity,[27] but was based on enduring relations and practice. Caravan roads are known already to have led from Tadmor to the Euphrates, linking it with the city of Kanesh in Anatolia and the kingdoms of Mari and Qatna in Syria in the Bronze Age.[28] The caravans in the overland routes were then initially donkey caravans[29] (Fig. 2.16); later, the camel became more common (Figs. 2.15, 2.18). Dromedaries are typical of the region and are seen grazing in the area today (Fig. 2.17).

In the 1930s, Michael Rostovtzeff paid special attention to the role of the caravan cities such as Palmyra, Dura Europos, Hatra, Jerash, and Petra in the East in antiquity. In his view, Palmyra was the foremost caravan city of them all.[30] Camels and caravans were means to bring all kinds of luxury commodities from faraway lands through Palmyra to the Mediterranean coast and Rome. Hatra was an important caravan city in the Persian frontier under the Parthians. It was a central site in the area of north-eastern Mesopotamia and, like Palmyra, was surrounded by pastoral nomads.[31] Compared to Hatra and Palmyra, the relationship between Palmyra and Dura Europos was even closer – they were like a sister and a brother – connected by a c. 200 km-long caravan road over the desert to the Euphrates. Unfortunately, the caravan cities of Hatra, Dura Europos, and Palmyra all fell into the hands of ISIS and faced destruction and lootings of their archaeological remains in 2014–2017. Either Persian or Roman conquests under various dynasties and rulers had taken place in the same cities in antiquity.

Rostovtzeff studied how ancient caravan cities formed a network through the Silk Road to the Roman world.[32] As a specialist in the Roman economy, he understood the importance of such trade networks that existed between the Orient and the Empire. He particularly studied

Fig. 2.15 *A donkey and dromedaries passing a necropolis in Palmyra in 1925–1946.* Photo: Library of Congress, Matson Collection

Fig. 2.16 *Onagers, donkeys, and mules were first used in caravaneering and as draft animals in the Near East. Animals eating near Halabiya at the edge of the ancient territory of Palmyra.* Photo: Minna Lönnqvist 2008, SYGIS

Fig. 2.17 *Dromedaries grazing in the desert surrounding Palmyra.* Photo: Minna Lönnqvist 2008, SYGIS

Dura Europos and excavated there, but he also published caravan inscriptions from Palmyra and identified the special gods that were dedicated to its caravan trade.[33] Common features in art and architecture appeared as cultural contacts flowed along the roads, and are visible

Fig. 2.18 *A soldier with an armour bearer and a camel. A relief from Palmyra, 2nd century* AD. Photo: Kenneth Lönnqvist 2004, SYGIS

in the line of caravan cities, especially in Roman times. Further to the south there was the caravan city of Petra near the Red Sea, now in southern Jordan, and the Persian influence was also felt there. Johann Ludwig Burkchardt identified architectural similarities between Palmyra and Petra.[34] Petra was on the spice and incense road of the Nabataean kingdom that stretched its borders inside present Syria and reached the Red Sea at Leuke Kome.[35]

Since Rostovtzeff several scholars have studied the ancient caravan roads of Palmyra. The late Khaled Al-As'ad was especially interested in them, but Michal Gawlikowski has also traced their possible lines. A special conference was organised in Syria on Palmyra and the Silk Road in the 1990s. Al-As'ad divided the direct trade routes from Palmyra into five branches or chains: 1) Palmyra–Damascus, 2) Palmyra–Homs, 3) Palmyra–Euphrates, 4) Palmyra–Isyria and 5) Palmyra–Apamea–Aleppo–Antioch.[36] The site where caravans stopped in Palmyra was in the Valley of the Tombs near the tower tomb of Elahbel (see Ch. 9). There the western entrance to the city was located. Khaled Al-As'ad and Jean-Baptiste Yon have assumed that the caravan commerce was not in the hands of the Palmyrenes but was directed from Tyre and Sidon, Phoenician cities on the Mediterranean coast.[37] There one could obtain precious purple dye and wines around the Mediterranean.

Eivind Seland has further studied the western route towards Antioch, the city that was also connected with Palmyra and illustrated in *Tabula Peutingeriana,* a 4th century AD Roman itinerary or a map.[38] One of the later travellers describes the arrival through the western road in the 19th century as follows:

> Nine p.m. We commenced a gradual ascent and by a bright moonlight arrived at two lofty ruined towers, and skirted along the mountain side by the edge of a deep valley thickly studded with ruined towers, the splendid burial monuments of the ancient Palmyrenes. Some along the side of the hill shone brightly in the full moonlight, while others spread their dark crumbling masses, shaded by projecting angle of the mountain, or dimly and shadowy in the gloom of the valley below.[39]

It is probable that the western trade concerned the Phoenician control or the control of Antioch, and that the desert roads were more in the hands of the Palmyrenes. Palmyra had special merchants and trade lords[40] who protected caravans on their overland routes. Besides Petra having control of the Red Sea trade and incense and spice routes,[41] Palmyra had a connection to the Persian Gulf and its maritime trade through the Euphrates (Figs. 2.19–2.21) at Spasinou Charax and Vologesias in Mesopotamia.[42] *Periplus Maris Erythraei*, a Greco-Roman source known as the 'Voyage Around the Erythraean Sea,' mentions ports and sea coasts used in antiquity in particular.[43] An honorific inscription associated with a statue in Palmyra is dedicated by a symposium of leather workers and floating skin makers to Septimius Hairan (Herod), Odenathus's son and stepson of Zenobia.[44] The

Fig. 2.19 *Major routes between Asia and Roman East in Late Antiquity (after Dura Europos on the Euphrates was destroyed). Source: R. Mouterde and A. Poidebard 1945.*

Fig. 2.20 *An ancient road and a bridge on the Euphrates in the region that was sometimes also under Palmyra.* Photo: Minna Lönnqvist 2005, SYGIS

Fig. 2.21 *An ancient boat carved in relief in stone and displayed at the Palmyra Museum recording the maritime connections of Roman Palmyra.* Photo: Kenneth Lönnqvist 2000, SYGIS, Courtesy of the Palmyra Museum

floating skins have been seen as a reference to skin rafts that were used on the Euphrates by the ancient Mesopotamians and indeed, have been used until modern times. A relief of a Roman boat in the Palmyra Museum refers to the Euphrates way and apparently to the sea trade of Palmyra (Fig. 2.21). Rome was interested in taxation of goods, and the Tax Law of Palmyra, found in the vicinity the Tariff Court of the city, was the way to control the trade.[45] We will further discuss this important document in association with the *Agora* and the Tariff Court in Palmyra (Ch. 9), sites near the area where caravans arrived through the Valley of Tombs.

Stars served as guides for the caravans navigating at night, and Venus especially was an important planet to be followed.[46] During the day, the movements of the sun were observed. No wonder the people of Palmyra focused their religion on the sun and astral beliefs in general. But caravans had various other means of navigating in the desert. There were *rujm* markers – piles of stones – serving as signposts or pathfinders. As previously mentioned, cairn/*tumulus* tombs also served this purpose in the deserts and steppes.[47]

Apart from Roman sites, Poidebard also prospected, photographed, and mapped caravan roads from the air.[48] Poidebard detected the possible ancient road between Palmyra and Hit on the Euphrates,[49] as Sir Aurel Stein did[50] (see Fig. 2.22). Poidebard also documented the alignment of the *Strata Diocletiana*, the Diocletian Road (Fig. 2.23), established by Emperor Diocletian (ruled AD 284–305) marking the eastern border zone of the Empire. It led from Damascus through Palmyra to the Euphrates. The *Strata* has been seen as the continuation of the earlier *Via Nova Traiana*, established by Emperor Trajan and running from Jordan to Bostra in Syria,[51] but the *Strata* continued further to the Middle Euphrates. This was the Roman eastern frontier zone that at the Euphrates continued the line of the river.

However, it has been suggested that during the rule of Emperor Trajan (ruled AD 98–117) (possibly already under Vespasian) a Roman road already connected Palmyra with the Euphrates.[52] This would accord with the sites mentioned in Ptolemy's Geography dating from the 2nd century AD[53], and a milestone, *miliarium*, associated with Trajan and found in the region (see Ch. 5, Fig. 5.6). It seems logical that this alignment was following oases and was earlier used as a caravan road. The alignment of the *Strata Diocletiana*, a dirt track, is visible in its milestones, various camps, forts and fortresses, *castri* and *castelli*, established to protect the frontier zone; some of these will be further discussed in the following chapters. According to

Fig. 2.22 *A caravan road from Palmyra to Hit photographed by Antoine Poidebard from air.*
Source: Poidebard 1934

Fig. 2.23 *A satellite map displaying the Strata Diocletiana leading through Palmyra (2) to the Euphrates. 1. Kh. al-Hallabat, 2. Palmyra, 3. 'Araq, 4. Helela, 5. Sukhne, 6. Taibe (Oriza), 7. El-Kowm, 8.Qdeir, 9. Tell al-Fhada, 10. Holle, 11. Resafa-Sergiopolis, 12. Qusair al-Saila, 13. Sura I and Sura II.* Courtesy: GORS

the milestones the *Strata* is dated c. AD 306.[54] Poidebard tried to identify in the *Strata* the military posts mentioned in *Notitia Dignitatum* that documents the administration of the Late Roman Empire; the original form of the text concerning the eastern parts dates from the late 4th century AD. [55]

Roads and tracks can be detected not only from the air but also from space with the help of modern technology using satellite photographs and images.[56] Hollow ways, old tracks, and roads have been traced with the CORONA satellite photographs.[57] As discussed, Landsat images, even at lower spatial resolution, reveal ancient and more modern caravan roads clearly in the desert landscape. Navigating by stars becomes obvious by looking at the images and the straight lines crossing deserts.[58] Many of the straight lines trodden in the desert sand radiate from the oasis of Palmyra in various directions (see Fig. 2.4).[59]

## 2.4 Textiles and colours on the Silk Road

Textiles were commodities that were traded by the people of the neighbourhood of Tadmor (Palmyra) in the Bronze Age. It is apparent that wool was a common material then, because of the pastoral nomads that produced it in the steppes.[60] As the name of the Silk Road indicates, the connection bringing silk from China to the Roman world started a real luxury trade. The secret of the mulberry tree was known in China, and fine quality silks were exported. Textiles were the source of wealth and pride of Palmyra that – as previously indicated – was one of the stations on the Silk Road network between China and the Roman Empire.[61] There were several branches of the Silk Road. Palmyra's wealth grew, especially in the 2nd century AD, along with the trade in textiles.[62]

Weaving has been associated with women, and spindle whorls made of ceramics are commonly found on ancient sites and are evidence that weaving once occurred in those places. Several female portraits from Palmyra show women holding a spindle whorl.[63] Weaving was considered a mythical act in antiquity, reminiscent of Mother Earth weaving the world together. We see this view of weaving as a sacred act in the Greco-Roman world: Penelope while waiting for Odysseus to return from the Trojan war to Ithaca concentrated on weaving.[64] From the ancient world we have special laws – edicts that provide information about the textile production in Late Antique and Byzantine worlds. The textiles, especially Damascene and Gobelin, have been sold in the markets of Palmyra and Damascus for hundreds of years (see Fig. 2.27).

The Palmyrenes certainly knew how to dress! We have been left with plentiful direct archaeological evidence of textiles from Palmyra and its region.[65] In addition, textile studies can be carried out using the presentation of garments in portraiture and other types of sculptures and paintings found in Palmyra (Fig. 2.24). Over 2,000 textile pieces have been preserved in Palmyra;[66] the dry desert climate has conserved them well. The greatest number of textiles from Palmyra comes from funerary contexts. Silk, wool, and linen were the main textiles in the Palmyrene trade during the Greco-Roman period, and these materials were used in clothes by the Palmyrenes themselves. Some cotton has also been found in the neighbourhood.

The wall paintings from Dura Europos illustrate a number of different costumes, and inscriptions from Dura even report prices of garments. As well as Greco-Roman clothes,

Fig. 2.24 *A funerary portrait oriental Palmyrene woman called Akma veiled and wearing plenty of jewellery, 2nd century AD.* Photo: Kenneth Lönnqvist 2008, courtesy of the Palmyra Museum

there are names of oriental clothing. Blankets, cushions, and bands are also included.[67] Palmyra followed Parthian and Greco-Roman styles in costume fashions, but it also developed its own sense of style. The Parthian influence could be seen in the use of trousers and short tunics (Fig. 2.25), unlike in the Greek dress called *chiton* and the Greek *himation* like a mantle, or Roman togas draped around and over a shoulder. These Greek and Roman types of clothing also appear in the art of Palmyra. Metal belts and embroidering were preferred, even metal threads common in the oriental style. Boots with trousers lavishly decorated with floral patterns were worn by men. Elaborate drapery is visible in a sculpture showing a Parthian costume with embroidered leggings and folds, like that seen in Buddhist styles (Fig. 2.25). Children have long dresses, or tunics and trousers – examples have been found in graves at Halabiya that belonged to the influence zone of Palmyra on the Euphrates.[68]

*Fig. 2.25 A sculptured male torso wearing clothing with elaborate drapery and embroidery in a Parthian style in Palmyra.* Photo: Kenneth Silver 2000, SYGIS, Courtesy of the Palmyra Museum

Chinese silks were of high quality in Palmyra, which preferred silks from China over India, and especially during the Han dynasty the connection through the Silk Road was a very active route bringing textiles to the Roman East. In Palmyra, Chinese silk was used from the 1st century BC to the 2nd century AD in particular, and as mentioned previously the 2nd century AD was the wealthiest period for Palmyra and its caravan trade. Chinese patterns of mythological animals have been found in ancient silk textiles in Palmyra, and silk was used as bands as well.[69] But the East and the West traditions were sometimes fused.[70] Such detailed Hellenistic patterns as grape pickers have been traced in the Chinese silks of Palmyra.[71] As well as imports, there were also local production and imitation in the silks of the region.[72]

In Dura Europos, in the textiles the same multicoloured four-petal flower is preferred as is seen in decoration at the Temple of the Palmyrene Gods and paintings of the synagogue

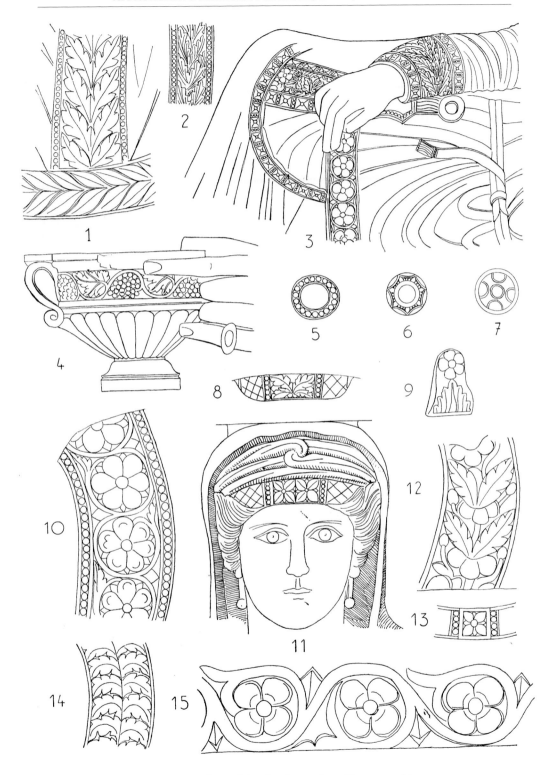

Fig. 2.26 *Sculptured details of vegetal motifs presented in costumes in one of the tombs of Palmyra.* Syria journal, 1936

and a Palmyrene house.[73] Several floral patterns have also been traced in the architecture of Palmyra; the same patterns were favoured both in textiles and architecture[74] (Fig. 2.26).

The Palmyrenes had local wool in abundance because of the mentioned availability, through pastoral nomads, of livestock and their products from sheep, lambs, goats, and camels grazing in the surrounding steppe. In Dura Europos the majority of textile finds are indeed wool.[75] The ancients used wool in their robes to protect them from cold; deserts at night can be bitterly cold. The Syrian wool was extraordinary high in quality during the 2nd and 3rd centuries AD. The tight weave is shown by the use of $26 \times 160$ threads for a square centimetre.[76] Wool mixed with camel hair was also used. Camel hair and goat hair are still utilised in tent and saddlebag weaving among the Bedouins of the region today. Linen appeared in various qualities. Egyptian linen, like that of Antioch, was of fine quality in antiquity, but Palmyra could have produced its own commodities as well. Gobelin flowers decorated linen tunics. Mummies that have been found in tower tombs in Palmyra were bound by linen bandages (see Ch. 9). Linen was used for bodies in the Roman East, as described in the burial of Jesus in the Bible.[77]

Various colours were used in textiles both in Palmyra and Dura Europos. Colours appear in bands and waves that were used to decorate clothes. Wool and linen could have been dyed with purple to add noble splendour to the textile, worth for the aristocracy. However, a white background seems to have been preferred in Dura, mixing linen and wool for decoration. The luxurious combination of purple decorated with golden thread has been found in Palmyra as well showing a difference with Dura.[78] The most famous red-purple dye was acquired from Phoenicia, especially Tyre, not far from Damascus on the Lebanese coast, where it was collected from the molluscs (*Murex*) collected from the Mediterranean Sea. The dyeing itself could have also taken place there. The Roman architect Vitruvius from the 1st century BC mentioned that purple exceeded all the colours both in costliness and its delightful effect. Vitruvius acknowledged that it can be found in other shades in the north, like black in Pontus and Gaul, but those nearest to the sun produced the best effects.[79]

*Historia Augusta, Life of Aurelian* (Part 2, 29, 1), an ancient source dealing with Palmyra of the 3rd century AD in the time of Zenobia, reports colours of the textiles in Rome:

> Concerning this I desire to say at least a few words. For you remember that there was in the Temple of Jupiter Best and Greatest on the Capitolium a short woollen cloak of a purple hue, by the side of which all other purple garments, brought by the matrons and by Aurelian himself, seemed to fade to the colour of ashes in comparison with its divine brilliance. This cloak, brought from the farthest Indies, the King of the Persians is said to have presented as a gift to Aurelian, writing as follows: 'Accept a purple robe, such as we ourselves use.' But this was untrue. For later both Aurelian and Probus and, most recently, Diocletian made most diligent search for this species of purple, sending out their most diligent agents, but even so it could not be found. But indeed it is said that the Indian sandyx yields this kind of purple if properly prepared.

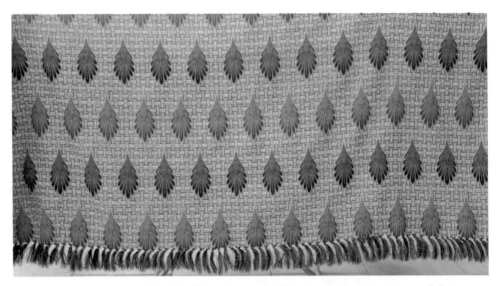

Fig. 2.27 *A modern table cloth of the Palmyrene style with palm leave patterns, a gift from Waleed al-As'ad to Minna Lönnqvist (Silver) delivered in 2005.* Photo: Minna Silver 2016

Generally, however, red for common people was acquired from dyer's madder, *Rubia tinctoria*. Indigo was also very popular.[80] Indigo for blue appears in the textiles, for example at Halabiya[81] that was under the influence of Palmyra. The source for indigo in the silks of Halabiya has been traced to China, but Palmyra may have obtained the colour from India. Palmyra had connections to the Indian trade through the Euphrates line that led to the Persian Gulf, from where pearls and precious stones were brought. However, common people did not use indigo in the Mediterranean region but another dye from woad *Isatis tinctoria*.[82]

Vitruvius further provides a detailed description of various sources from the natural world and artificially produced substances for other colours and their properties. Nature provides yellow ochre, red earths, Melian white, and green chalk. Cinnabar and quicksilver were used for producing artificial gold. Blue was, according to Vitruvius, invented in Alexandria, from sand, copper, and flowers of natron in an earthenware vessel put into an oven. He further explains how black was made for ink by burning resin using a certain method.[83] Lapis lazuli, the blue stone favoured in the east, had its sources in the mountains of Afghanistan.[84] So antiquity was colourful, and the white marble and limestone visible in ancient buildings and portraiture do not provide us today with the right image of the past. When we look at those ruins and remnants the time of antiquity looks pale and colourless. Only in some textiles, frescoes, and mosaics do we achieve the right impression of the splendour that has preserved down to us.

The Diocletian edict fixed the maximum prices for the textiles in the Late Roman Empire in the early 4th century AD.[85] Raw silk and wool were priced – the washed wool could be fine and expensive or just ordinary and cheaper. The prices of women's and men's shoes were also specified.

# Endnotes

1   Poidebard, A. (1934) *La Trace de Rome dans le Désert de Syrie: Le Limes de Trajan à la conquête arabe. Recherches aériennes (1925–1932)*. Texte. Atlas. BAH. Tome XVIII. Paris: Paul Geuthner.

2   See Gregory, S. and Kennedy, D. (1985) *Sir Aurel Stein's Limes Report*. BAR International Series 272 (i-ii). Oxford: B.A.R.

3   Rees, L.W.B. (1929) The Transjordan Desert, in *Antiquity*, Vol. III, pp. 389–407.

4   Brooks, R.R. and Johannes, D. (1990) *Phytoarchaeology*, Historical, Ethno- & Economic Botany Series, Vol. 3, T.R. Dudley, General Editor, Portland: Oregon: Dioscorides Press, pp. 133, 166 & 167. Poidebard, A. (1934) *La Trace de Rome dans le Désert de Syrie: Le Limes de Trajan a la conquête arabe. Recherches aériennes (1925–1932)*. Texte. Atlas. BAH. Tome XVIII. Paris: Paul Geuthner.

5   Evans, J. and O'Connor, T. (1999) *Environmental Archaeology, Principles and Methods*. Midsomer Norton, Somerset: Sutton Publishing, p. 78.

6   Kennedy, D. (1998) Declassified satellite photographs and archaeology in the Middle East: case studies from Turkey, in *Antiquity*, Vol. 72, pp. 553–561; Lönnqvist, M. and Törmä, M. (2003) SYGIS – The Finnish Archaeological Project in Syria, in Proceedings of the XIXth International Symposium CIPA 2003 (The ICOMOS & ISPRS Committee for Documentation of Cultural Heritage: New Perspectives to Save Cultural Heritage, Antalya, Turkey), ed. by Altan, M.O., in *The ISPRS International Archives of the Photogrammetry, Remote Sensing and Spatial Information Sciences*, Vol. XXXIV-5/C15, pp. 609–614.

7   Lönnqvist, M., Törmä, M., Lönnqvist, K. and Nuñez, M. (2014) Highland – Lowland Human Interaction in Ancient Syria, in *The ISPRS International Archives of Photogrammetry, Remote Sensing and Spatial Information Sciences*, Vol. XXXIXB4, XXII ISPRS Congress 25 August – 01 September 2012, ed. Madden, M. and Shortis, M., Melbourne, Australia, Göttingen: Copernicus Publications, pp. 455–460; Silver, K., Silver, M., Törmä, M., Okkonen, J. and, T. Okkonen (2017) Applying Satellite Data Sources in the Documentation and Landscape Modelling for Graeco-Roman/Byzantine Fortified Sites in the Tur Abdin Area, Eastern Turkey, in *ISPRS Annals of the Photogrammetry, Remote Sensing and Spatial Information Sciences*, Volume IV-2/W2, 2017, 26th International CIPA Symposium 2017, 28 August–01 September 2017, Ottawa, Canada, pp. 251–258: https://doi.org/10.5194/isprs-annals-IV-2-W2-251-2017.

8   See Silver, M. (2016) Conservation Techniques in Cultural Heritage, in *3D Recording, Documentation and Management of Cultural Heritage*, ed. by Stylianidis, E. and Remondino, F., Caithness, Scotland: Whittles Publishing, pp. 15–105, especially p. 29.

9   Le Tensorer, J.-M., Jagher, R, Rentzel, P., Hauck, T., Ismail-Meyer, K., Pümpin, C. and Wojitczac, D. (2007) Long-term Site Formation Processes at the Natural Spring Nadaouiyeh and Hummal in the El Kowm Oasis, Central Syria, in *Geoarchaeology*, Vol. 22, pp. 621–639.

10  See Akazawa, T. (1996) A Reconstruction of the Middle Palaeolithic Cultural Ecology of the Douara Cave, Palmyra, Central Syria, in *Les Annales Archéologiques Arabes Syriennes*, Vol. XLII, 1996, *Special Issue Documenting the Activities of the International Colloquium Palmyra and the Silk Road*, pp. 63–73.

11  Legge, A.J. and Rowley-Conwy, P.A. (1987) Gazelle Killing in Stone Age Syria, in *Scientific American*, 257, pp. 76–83.

12  Legge, A.J. and Rowley-Conwy, P.A. (1987) Gazelle Killing in Stone Age Syria, in *Scientific American*, 257, pp. 76–83.

13  Lönnqvist, M. (2000) *Between Nomadism and Sedentism: Amorites from the Perspective of Contextual Archaeology*, Helsinki: Juutiprint, p. 90; Morandi Bonacossi, D. (2014) Desert-kites in an Aridifying Environment: Specialised Hunter Communities in the Palmyra Steppe during the Middle and Late Holocene, in *Settlement Dynamics and Human-Landscape Interaction in the Dry Steppes of Syria*, ed. by Morandi Bonacossi, D., Wiesbaden: Harrassowitz Verlag, pp. 33–47.

14  Rees, L.W.B. (1929) The Transjordan Desert, in *Antiquity*, Vo. III, pp. 389–407.

15  Crassard, R., Barge, O., Bichot, C.-E., Brochier, J. E., Chaoud, J., Chambrade, M.-L., Chataigner, C., Madi, K., Regagnon, E., Hamida, S., and Vila, E. (2015) Addressing the Desert Kites Phenomenon and Its Global Range Through a Multiproxy Approach, in *Journal of Archaeological Method and Theory*, Vol. 22, pp. 1093–1121.

16  Burckhardt, J. L. (1831) *Notes on the Bedouins and the Wahabys collected during the travels by late John Lewis Burckhardt*, Vol. I, London: Henry Colburn and Richard Bentley, p. 220.

17  Boëda, E., Geneste, J.M., Griggo, C., Mercier, N., Muhesen, S., Reyss, J.L., Taha, A. and Valladas, H. (1999) A Levallois point embedded in the vertebra of a wild ass (Equus Africanus): hafting, projectiles and Mousterian hunting weapons, in *Antiquity*, Vol. 73, No. 280, pp. 394–402.

18  Owen, J. (2006) Extinct "Elephant Size" Camel Found in Syria, in *National Geographic*, October 11.

19  Lönnqvist, M., Törmä, M., Lönnqvist, K. and Nuñez, M. (2011) *Jebel Bishri in Focus: Remote sensing, Archaeological surveying, mapping and GIS studies of Jebel Bishri in central Syria by the Finnish project SYGIS*, BAR International Series 2230, pp. 107–108.

20  Moore, A., Hillman, G.C. and Legge, A.J. (2000) *Village on the Euphrates: From Foraging to Farming at Abu Hureyra*. Oxford: Oxford University Press.

21  Perkins, D. Jr. (1964) Prehistoric Fauna from Shanidar, Iraq, in *Science*, Vol. 144, pp. 1565–1566.

22  Zeder, M. and Hesse, B. (2000) The Initial Domestication of Goats (Capra hircus) in the Zagros Mountains 10,000 Years Ago, in *Science*, Vol. 287, pp. 2254–2256.

23  Stordeur, D. (1993) Sédentaires et nomades du PPNB final dans le desert de Palmyre (Syrie), in *Paléorient*, Vol. 19, pp. 187–204; Stordeur, D. and Taha, A. (1996) Ressemblances et dissemblances entre les sites nomades et sédentaires de la steppe syrienne au VIe millénaire, in *Les Annales Archéologiques Arabes Syriennes*, Vol. XLII, 1996, *Special Issue Documenting the Activities of the International Colloquium Palmyra and the Silk Road*, pp. 85–98; Zarins, J. (1989) Jebel Bishri and the Amorite Homeland: the PPNB Phase, in *To the Euphrates and Beyond*, Archaeological Studies in Honour of Maurits N. Van Loon, ed. by Chaex, O.M. et al., Rotterdam: Balkema, pp. 29–52; Zarins, J. (1992) Archaeological and Chronological Problems within the Great Southwest Asian Arid Zone, 8500–1850 B.C., in *Chronologies in Old World Archaeology*, Third Edition, 2 Vols., ed. by Ehrich, R.W., Chicago: The University of Chicago Press, pp. 42–62; see datings by J. Cauvin in Anfinset, N. (2008) Towards the Specialised Food Production – A Look at Jebel Bishri, in *Jebel Bishri in Context: Introduction to the Archaeological Studies of Jebel Bishri and its Neighbourhood in Central Syria*, Proceedings of a Nordic Research Training Seminar in Syria, May 2004, ed. by Lönnqvist, M., in BAR International Series 1817, Oxford: Archaeopress, pp. 1–13.

24  Lönnqvist, M. (2014) The Emergence of Pastoral Transhumance at Jebel Bishri in Central Syria, in *Settlement Dynamics and Human-Landscape Interaction in the Dry Steppes of Syria*, ed. by Morandi Bonacossi, D., Wiesbaden: Harrassowitz Verlag, pp. 93–109.

25  Iamoni, M. (2014) Late Neolithic Funerary Evidence from Palmyra: the Rujem al-Majdur necropolis and the "desert-kite and tumulus cultural horizons" in Central Syria, in *Settlement Dynamics and Human-Landscape Interaction in the Dry Steppes of Syria*, ed. by Morandi Bonacossi, Daniele, Studia Chaburensia Vol. 4, Wiesbaden: Harrassowitz Verlag, pp. 49–61.

26  Silver, M. (2016) Visualizing Invisible Nomads, in *World Heritage Strategy Forum*, 9–11 September, 2016, the Loeb House, Harvard University.

27  See Joannés, F. (1997) Palmyre et les Routes du Desert au debut de Déuxieme Millenaire av. J.-C., in *M.A.R.I., Mari Annales de Recherches Interdisciplinaires* 8, Paris: Éditions Recherche sur les Civilisations, pp. 393–415.

28  See Klengel, H. (1996) Palmyra and International Trade in the Bronze Age: the Historical Background, in *Les Annales Archéologiques Arabes Syriennes*, Vol. XLII, 1996, *Special Issue Documenting the Activities of the International Colloquium Palmyra and the Silk Road*, pp. 159–163; Joannés, F. (1997) Palmyre et les Routes du Desert au debut de Déuxieme Millenaire av. J.-C., in *M.A.R.I., Mari Annales de Recherches Interdisciplinaires* 8, Paris: Éditions Recherche sur les Civilisations, pp. 393–415.

29  See Lewy, J. (1952) Studies in the Historic Geography of the Ancient Near East, Old Assyrian Caravan Roads in the Valley of Habur and the Euphrates and in Northern Syria, in *Orientalia*, Vol. XXI, pp. 339–425.

30  Rostovtzeff, M. (1932) Repr. 1971. *Caravan Cities*. New York: AMS Press.

31  See, e.g., Hauser, S.R. (2000) Ecological Limits and Political Frontiers: The "Kingdom of the Arabs" in the Eastern Jazira in the Arsacid Period, in *Landscapes: Territories, Frontiers and Horizons in the Ancient Near East, Part II, Geography and Cultural landscapes, Papers presented to the XLIV Rencôntre Assyriologique Internationale, Venezia, 7-11 July 1997*, ed. Milano, L., de Martino, S., Fales, F.M. and Lanfranchi, G.B., History of the Ancient Near East/Monographs–III/2, Padova: Sargon srl, pp. 187–201.

32  Rostovtzeff, M. (1932) Repr. 1971. *Caravan Cities*. New York: AMS Press.

33  Rostovtzeff, M. (1932) Les inscriptions des caravanières de Palmyra, in *Mélanges Gustave Glotz*, Vol. II, Paris: Presses Universitaires de France, pp. 793–811 and Rostovtzeff, M. I. (1932) The Caravan-Gods of Palmyra, in *The Journal of Roman Studies*, Vol. 22, pp. 107–116

34  See Burckhardt, J.L. (1822) *Travels in Syria and the Holy Land*, London: John Murray, pp. 422–433.

35  See Stoneman, R. (1995) *Palmyra and Its Empire: Zenobia's Revolt against Rome*, Ann Arbor: The University of Michigan Press, pp. 34–38; see also Fiema, Z. (1996) Nabatean and Palmyrene Commerce – the Mechanisms of Intensification, in *Les Annales Archéologiques Arabes Syriennes*, Vol. XLII, 1996, *Special Issue Documenting the Activities of the International Colloquium Palmyra and the Silk Road*, pp. 191–195.

36  See Al As'ad, Kh. (1996) Caravan Roads of Palmyra, in *Les Annales Archéologiques Arabes Syriennes*, Vol. XLII, 1996, *Special Issue Documenting the Activities of the International Colloquium Palmyra and the Silk Road*, pp. 123–124; Gawlikowski, M. (1996) Palmyra and its Caravan Trade, in *Les Annales Archéologiques Arabes*

*Syriennes*, Vol. XLII, 1996, *Special Issue Documenting the Activities of the International Colloquium Palmyra and the Silk Road*, pp. 139–145.

37  As'ad, Kh. and Yon, J.-B. (2001) *Inscriptions de Palmyre, Promenades épigraphiques dans la ville antique de Palmyre,* Guides archéologiques de l'Institut Français d' Archéologie du Proche-orient No. 3, Beyrouth-Amman-Damas: l'Institut Français d'Archéologie du Proche-Orient, p. 78.

38  Seland, E. (2008) Trade Routes of Palmyra, with Special Notes on Western Routes in the Palmyrene Trade, in *Jebel Bishri in Context: Introduction to the Archaeological Studies and the Neighbourhood of Jebel Bishri in Central Syria*, Proceedings of a Nordic Research Training Seminar in Syria, May 2004, ed. by Lönnqvist, M., BAR International Series 1817, Oxford, Archaeopress, pp. 89–97.

39  Greenstreet Addison, C. (1838) *Damascus and Palmyra: a journey to the East with a sketch of the State and Prospects of Syria under Ibrahim Pasha.* Vol. 1. London. Richard Bentley.

40  Gregoratti, L. (2016) Palmyrene Trade Lords and Protections of the Caravans, in *ARAM,* Vol. 27: 1 & 2, 2015, pp. 139–148.

41  Stoneman, R. (1995) *Palmyra and Its Empire, Zenobia's Revolt against Rome*, Ann Arbor: The University of Michigan Press, pp. 32–41, see Fig. 2 on p. 41, the map on the caravan roads.

42  Gawlikowski, M. (1996) Palmyra and its Caravan Trade, in *Les Annales Archéologiques Arabes Syriennes*, Vol. XLII, 1996, *Special Issue Documenting the Activities of the International Colloquium Palmyra and the Silk Road*, pp. 139–145, especially pp. 140–141.

43  See, for example, W.H. Schoff (1912) *The Periplus of the Erythraean Sea: Travel and Trade in the Indian Ocean by a Merchant of the First Century.* New York: Longmans.

44  Al As'ad, Kh. and Gawlikowski, M. (1997) *The Inscriptions in the Museum of Palmyra, A Catalogue,* Palmyra & Warsaw: The Committee for Scientific Research, p. 23.

45  See Matthews, J.F. (1984) The Tax Law of Palmyra: Evidence for Economical history in a City of the Roman East, in *The Journal of Roman Studies*, Vol. 74, pp. 157–180; see also Lönnqvist, K. (2008) The Tax Law of Palmyra and the Introduction of the Roman Monetary System to Syria – A Re-evaluation, in *Jebel Bishri in Context: Introduction to the Archaeological Studies and the Neighbourhood of Jebel Bishri in Central Syria*, Proceedings of a Nordic Research Training Seminar in Syria, May 2004, ed. Lönnqvist, M., BAR International Series 1817, pp. 73–88.

46  Madoun, M.A. (1996) *The Silk Road and Palmyra: Bartering Cultures*, Damascus: Al Namir Publishing House, pp. 69–70.

47  Lönnqvist, M., Törmä, M., Lönnqvist, K. and Nuñez, M. (2011) *Jebel Bishri in Focus: Remote sensing, Archaeological surveying, mapping and GIS studies of Jebel Bishri in central Syria by the Finnish project SYGIS*, BAR International Series 2230, Oxford: Archaeopress, p. 350; see also Madoun, M.A. (1996) *The Silk Road and Palmyra: Bartering Cultures*. Damascus: Al Namir Publishing House.

48  See Poidebard, A. (1934) *La Trace de Rome dans le Désert de Syrie: Le Limes de Trajan a la conquête arabe: Recherches aériennes (1925–1932)*. Texte. Atlas. BAH. Tome XVIII. Paris: Paul Geuthner.

49  See Poidebard, A. (1934) *La Trace de Rome dans le Désert de Syrie: Le Limes de Trajan a la conquête arabe. Recherches aériennes (1925–1932)*. Texte. Atlas. BAH. Tome XVIII. Paris: Paul Geuthner.

50  See Gregory, S. and Kennedy, D. (1985) *Sir Aurel Stein's Limes Report*, BAR International Series 272 (i), Oxford: B.A.R., pp. 183–195.

51  See Borstad, K. (2008) History from Geography: The Initial Route of the Via Nova Traiana, in *Levant*, Vol. 40, pp. 55–70.

52  Edwell, P. (2008) *Between Rome and Persia: the middle Euphrates, Mesopotamia and Palmyra under Roman Control*, London and New York: Routledge, pp. 20–24.

53  Claudius Ptolemy, *The Geography*, Locations in Syria, Book V, transl. and ed. by Stevenson, E.L., New York: Dover Publications (1991).

54  See Lönnqvist, M., Törmä, M., Lönnqvist, K. and Nuñez, M. (2011) *Jebel Bishri in Focus: Remote sensing, Archaeological surveying, mapping and GIS studies of Jebel Bishri in central Syria by the Finnish project SYGIS*, BAR International Series 2230, Oxford: Archaeopress, p. 250.

55  See Seeck, O. (1876) *Notitia dignitatum, Accedunt notitia urbis Constantinopolitanae et laterculi provinciarum.* Berolini: Weidmann.

56  See Lönnqvist, M., Törmä, M., Lönnqvist, K. and Nuñez, M. (2011) *Jebel Bishri in Focus: Remote sensing, Archaeological surveying, mapping and GIS studies of Jebel Bishri in central Syria by the Finnish project SYGIS*, BAR International Series 2230, Oxford: Archaeopress, pp. 324–338. Also Silver, M., Törmä, M., Silver, K., Okkonen, J. and Nuñez, M. (2015) Remote Sensing, Landscape and Archaeology: Tracing Ancient Tracks and Roads between Palmyra and the Euphrates in Syria, in *ISRPS Annals of Photogrammetry, Remote Sensing and Spatial information Sciences*, Vol. II-5, W 3, 25th International CIPA Symposium 2015, 31 August – 04

September 2015, Taipei, Taiwan, pp. 279–285: http://www.isprs-ann-photogramm-remote-sens-spatial-inf-sci.net/II-5-W3/279/2015/isprsannals-II-5-W3-279-2015.pdf

57  Ur, J. (2003) CORONA Satellite Photography and Ancient Road Networks, in *Antiquity*, Vol. 77, pp. 102–115.

58  See Lönnqvist, M., Törmä, M., Lönnqvist, K. and Nuñez, M. (2011) *Jebel Bishri in Focus: Remote sensing, Archaeological surveying, mapping and GIS studies of Jebel Bishri in central Syria by the Finnish project SYGIS*, BAR International Series 2230, Oxford: Archaeopress, pp. 343–344.

59  See Silver, M., Törmä, M., Silver, K., Okkonen, J. and Nuñez, M. (2015) Remote Sensing, Landscape and Archaeology: Tracing Ancient Tracks and Roads between Palmyra and the Euphrates in Syria, in *ISRPS Annals of Photogrammetry, Remote Sensing and Spatial information Sciences*, Vol. II-5, W 3, 25th International CIPA Symposium 2015, 31 August – 04 September 2015, Taipei, Taiwan, pp. 279–285. http://www.isprs-ann-photogramm-remote-sens-spatial-inf-sci.net/II-5-W3/279/2015/isprsannals-II-5-W3-279-2015.pdf

60  Heltzer, M. (1981) *The Suteans*, with a contribution by Arbeli, S., Istituto Universitario Orientale, Seminario di Studi Asiatici, Series Minor XIII, Naples: Istituto Univeritario Orientale, p. 25; see especially Ziegler, N. (2014) The Sutean Nomads in the Mari Period, in *Settlement Dynamics and Human-Landscape Interaction in the Dry Steppes of Syria*, Studia Chaburensia, Vol. 4, ed. by Morandi Bonacossi, D., pp. 209–226.

61  Pfister, R. (1934) *Textiles de Palmyre*. Découverts par le Service des antiquités du Haut Commissariat de la R.F. dans la nécropole de Palmyre. Paris: Éditions d'art et d'histoire.

62  Rostovtzeff, M. (1932) Repr. 1971. *Caravan Cities*. New York: AMS Press; Will. E. (1996) Palmyrenes et les Routes de la Soie, in *Les Annales Archéologiques Arabes Syriennes*, Vol. XLII, 1996, *Special Issue Documenting the Activities of the International Colloquium Palmyra and the Silk Road*, pp. 125–128.

63  Finlayson, C. (2002) The Women of Palmyra – Textile Workshops and the Influence of the Silk Trade in Roman Syria, in *Textile Society of America Symposium Proceedings*, University of Nebraska – Lincoln Papers 385, pp. 2–5; see Sadurska, A. and Bounni, A. (1994) *Les sculptures funéraires de Palmyre en collaboration avec Khaled al-Ass'ad et Krzysztof Makowski*, Rivista di Archeologia, diretta da Traversari, G., Roma: Giorgio Bretschneider Editore, Figs. 131–142.

64  See Homer, *The Odyssey*, Book 2.

65  Pfister, R. (1934) *Textiles de Palmyre*. Découverts par le Service des antiquités du Haut Commissariat de la R.F. dans la nécropole de Palmyre. Paris: Éditions d'art et d'histoire; Schmidt-Colinet, A., Stauffer, A., Al As'ad, K. (2000) *Die Textilien aus Palmyra: Neue und Alte Funde*. Damaszener Forschungen 8. Mainz am Rhein: Philipp von Zabern.

66  Schmidt-Colinet, A., Al-As'ad, K. and Al-As'ad, W. (2016) Palmyra, 30 Years of Syro-German/Austrian Archaeological Research (Homs), in *A History of Syria in One Hundred Sites*, ed. by Kanjou, Y. and Tsuneki, A., Oxford: Archaeopress, pp. 339–348.

67  Pfister, R. and Bellinger, L. (1945) The Textiles, Part II, in *The Excavations at Dura-Europos*, conducted by Yale University and the French Academy of Inscriptions and Letters, Final Report IV, ed. by Rostovtzeff, M., Bellinger, A.R., Brown, F.E., Toll, N.P. and Welles, C.B., New Haven: Yale University Press, pp. 10–14.

68  Pfister, R. (1952) *Textiles de Halabiyeh (Zenobia)*: découverts par la Service de Antiquités de la Syrie dans la nécropole de Halabiyeh sur l'Euphrate. BAH XLVIII. Paris: Geuthner.

69  Pfister, R. (1934) *Textiles de Palmyre*. Découverts par le Service des antiquités du Haut Commissariat de la R.F. dans la nécropole de Palmyre. Paris: Éditions d'art et d'histoire; Pfister, R. and Bellinger, L. (1945) The Textiles, Part II, in *The Excavations at Dura-Europos*, conducted by Yale University and the French Academy of Inscriptions and Letters, Final Report IV, ed. by Rostovtzeff, M., Bellinger, A.R., Brown, F.E., Toll, N.P. and Welles, C.B., New Haven: Yale University Press, p. 3; Finlayson, C. (2002) The Women of Palmyra – Textile Workshops and the Influence of the Silk Trade in Roman Syria, in *Textile Society of America Symposium Proceedings, University of Nebraska – Lincoln Papers* 385, pp. 2–5.

70  Schmidt-Colinet, A. (1996) East and West in Palmyrene Pattern Books, in *Les Annales Archéologiques Arabes Syriennes*, Vol. XLII, 1996, *Special Issue Documenting the Activities of the International Colloquium Palmyra and the Silk Road*, pp. 420–423.

71  Żuchowska, M. (2014) "Grape Picking" Silk from Palmyra: A Han Dynasty Chinese Textile with a Hellenistic Decoration Motif, in *Światowit, Annual of the Institute of Archaeology of the University of Warsaw*, Vol. XII, Fasc. A, Mediterranean and Non-European Archaeology, pp. 143–161.

72  Stauffer, A. (1996) Textiles from Palmyra: Local Production and the Import and Imitation of Chinese Silk Weaving, in *Les Annales Archéologiques Arabes Syriennes*, Vol. XLII, 1996, *Special Issue Documenting the Activities of the International Colloquium Palmyra and the Silk Road*, pp. 425–430.

73  Pfister, R. and Bellinger, L. (1945) The Textiles, Part II, in *The Excavations at Dura-Europos*, conducted by Yale University and the French Academy of Inscriptions and Letters, Final Report IV, ed. by Rostovtzeff, M., Bellinger, A.R., Brown, F.E., Toll, N.P. and Welles, C.B., New Haven: Yale University Press, p. 8.

74  Schmidt-Colinet, A., Al-As'ad, K. and Al-As'ad, W. (2016) Palmyra, 30 Years of Syro-German/Austrian Archaeological Research (Homs), in *A History of Syria in One Hundred Sites*, ed. by Kanjou, Y. and Tsuneki, A., Oxford: Archaeopress, pp. 339–348, pp. 341–342.

75  Pfister, R. and Bellinger, L. (1945) The Textiles, Part II, in *The Excavations at Dura-Europos* conducted by Yale University and the French Academy of Inscriptions and Letters, Final Report IV, ed. by Rostovtzeff, M., Bellinger, A.R., Brown, F.E., Toll, N.P. and Welles, C.B., New Haven: Yale University Press, p. 1.

76  Nockert, M. (1989) Vid Sidenvägens ände: Textilen från Palmyra till Birka, in *Palmyra: Öknens drottning*, Stockholm: Medehavsmuseet & Statens historiska museum, pp. 77–105.

77  See Luke 24: 12.

78  Pfister, R. and Bellinger, L. (1945) The Textiles, Part II, in *The Excavations at Dura-Europos* conducted by Yale University and the French Academy of Inscriptions and Letters, Final Report IV, ed. by Rostovtzeff, M., Bellinger, A.R., Brown, F.E., Toll, N.P. and Welles, C.B., New Haven: Yale University Press, pp. 6–15.

79  Vitruvius, *The Ten Books on Architecture*, Book VII, Ch. XIII, XIV.

80  Pfister, R. and Bellinger, L. (1945) The Textiles, Part II, in *The Excavations at Dura-Europos,* conducted by Yale University and the French Academy of Inscriptions and Letters, Final Report IV, ed. by Rostovzeff, M., Bellinger, A.R., Brown, F.E., Toll, N.P. and Welles, C.B., New Haven: Yale University Press, pp. 6–15.

81  Pfister, R. (1951) *Textiles de Halabiyeh (Zenobia)*, découverts par le Service des Antiquités de la Syrie dans la Nécropole de Halabiyeh sur l'Euphrate, Paris: Libraire Orientaliste Paul Geuthner, p. 54.

82  Pfister, R. and Bellinger, L. (1945) The Textiles, Part II, in *The Excavations at Dura-Europos,* conducted by Yale University and the French Academy of Inscriptions and Letters, Final Report IV, ed. by Rostovtzeff, M., Bellinger, A.R., Brown, F.E., Toll, N.P. and Welles, C.B., New Haven: Yale University Press, p. 4.

83  Vitruvius, *The Ten Books on Architecture*, Book VII, Ch. VII–XIV.

84  See Majidzadeh, Y. (1982) Lapis lazuli and the Great Khorasan Road, in *Paléorient*, Vol. 8, pp. 59–69.

85  Graser, E.R. (1940) The Edict of Diocletian on Maximum Prices, in An Economic Survey of Ancient Rome, Vol. V: Rome and Italy of the Empire, ed. by Frank, T., Baltimore: Johns Hopkins Press, pp. 305–421.

# 3

# REVEALING CITIES BURIED BENEATH CITIES

## 3.1 Digging up the cycle of sedentary life

Fig. 3.1 *Tell Tibne on the Euphrates frontier zone of the Palmyrene territory neighbouring the fortress of Zenobia on Halabiya. Tibne has been identified with Diocletian's fortress of Mambri.* Photo: Minna Silver 2006, SYGIS

In the Ancient Near East there are thousands of mounds called *tell* in Arabic (*tepe* in Persian, *höyük* in Turkish, *tel* in Hebrew) that hide ancient villages, towns, or cities within their layers, one above another (see Tell Tibne in the frontier zone of the territory of Palmyra on the Euphrates Figs. 3.1, 3.2, and 3.3). They provide evidence of societies that once existed and are for archaeologists to be searched for and studied. Each layer has its history, style of architecture, and types of objects. The cycle of life, ageing, and ruination produces layers of deposits that may witness the past over thousands of years. Archaeologists are specialists in this cycle.

The first steps towards urbanisation in the Near East took place in the Neolithic period c. 8,000 BC. Jericho, situated in the desert oasis in the Jordan Valley,[1] now in the Palestinian territories of Israel, is an example of an early settlement with urban features. The somewhat younger Neolithic site of Çatal Höyük in Anatolia in modern Turkey is described by its excavator as a kind of a proto-city,[2] well-organised with its layout and evidence of specialisation. After these initial endeavours, explosive urbanisation took place in the Early Bronze Age in the Near East from c. 3,300 BC, especially in the great river valleys or oases. Sites in the river valleys and desert oases, such as Palmyra, situated below the 200–250 mm annual rainfall line were largely dependent on irrigated agriculture.[3] Rain-fed areas lie above the 200 mm annual precipitation. The area of uncertainty is defined as between the 200 mm and 320 mm isohyets. This so-called 'fragile Fertile Crescent' extends from Egypt to the Levant and the Jezira between the Twin Rivers. It is typically an agro-pastoralist region,[4] while desert and steppic areas are more nomadic areas.

The desert surrounding Palmyra has preserved ancient sites well, even on the surface, including the ruins of the ancient city of Palmyra itself. Archaeologists have also excavated ancient cities beneath these visible surfaces, and parts of Palmyra from the Bronze Age (3300–1200 BC), Hellenistic (330 BC–31/27 BC), and Roman period (27 BC–AD 395/476) have been revealed. Pottery has traditionally been used as the backbone of archaeological dating, and it still forms the greater bulk of archaeological source material in the region. Pottery types follow the taste of a certain culture and time, and they can be associated with them and used as relative

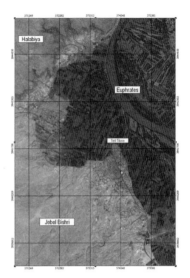

Fig. 3.2 *Mapping Tel Tibne on Landsat-7 by Minna Lönnqvist, SYGIS*

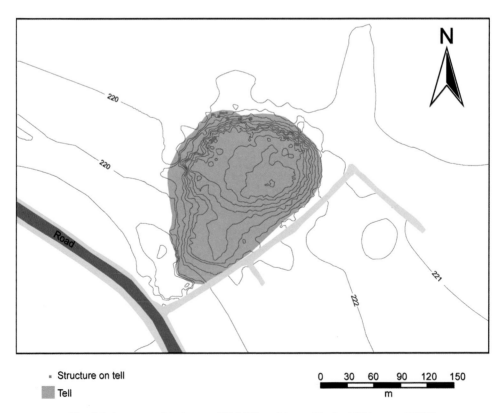

- Structure on tell

▨ Tell

0  30  60  90  120  150
m

Fig. 3.3 *A topographical map of Tell Tibne.* Mapped by Jari Okkonen, SYGIS

Fig. 3.4 *Studying washed pottery at the Guest House for archaeologists inside the precinct of the Temple of Bel in Palmyra.* Photo: Minna Lönnqvist 2005, SYGIS

dating material. They can indicate contacts and networks to the outside world. The layouts of cities, architectural styles, artistic objects like statues and paintings, as well as tombs and burial customs also identify the age of cultures. Inscriptions on stone, metal, pottery/terracotta, glass, texts on papyri and parchments can mediate information about precise dates. Seals, stamps, and coins also often can provide absolute dating as they bear information on historical rulers and therefore provide dates according to their tenure. Now, modern scientific methods, based on physics and chemistry, including radiocarbon dating, are used to refine the dating in calendric years. However, the radiocarbon method provides dating with a marginal error varying from decades to a few hundred years, and the results are dependent on calibrations based on tree-ring series provided by dendrochronological analyses.

In antiquity the Efqa spring was holy to the Palmyrene people because of its life-giving powers to the oasis and its plants. Altars surrounding the spring were dedicated to various deities. Therefore, it was guarded by gods that were important to Palmyra. One was nameless,[5] but Yarhibol in particular was an oracular god of the Efqa spring, and the god was identified with the sun.[6] Altars and symbols on other finds mediate the importance of the spring. Because of the presence of sulphates the spring provided healing properties as well. Now the spring is situated some metres beneath the present level of the city, showing how the building activities and accumulation of soil have raised the level of the city, and it has largely dried up. Because the annual precipitation in the area of Palmyra is less than 200 mm, it does not sustain cultivation without irrigation. An elaborate water system was used in the city in antiquity including springs, wells, cisterns, channels, and covered pipes. Water pipes

Fig. 3.5 *A covered Roman terracotta water pipe in Palmyra.* Photo: Kenneth Lönnqvist 2004, SYGIS

Fig. 3.6 *A Roman dam/barrage and a stepped water reservoir on Jebel Bishri in the Palmyrides.* Photo: Kenneth Lönnqvist 2003, SYGIS

from the Roman period are still visible in the city, demonstrating an elaborate water system (Fig. 3.5). Evidence of such was also found in the excavations of the Hellenistic town.[7] Water channels convey water into various parts of the city, and the underground networks reached 600–800 metres in ancient times.[8]

The Palmyrenes have been skillful in using water harvesting systems with dams, barrages and reservoirs (Fig. 3.6) in the surrounding desert-steppe areas. The Harbaqa dam is an immense constructional achievement that took place during the Roman period c.70 km southwest of Palmyra (Fig. 3.7).[9]

Ancient wells (Fig. 3.8), channels, or canals, known as *foggaras* or *qanats*, (Fig. 3.9) were used in channelling the water in the desert-steppe, some of which were identified and documented by the previously mentioned SYGIS project led by the present principal author.[10] More recent water systems in the district of Palmyra are seen in Figs. 3.10 and 3.11.

## 3.2 Bronze and Iron Age Tadmor

The Temple of Bel area in Palmyra is built on a *tell* and has provided information on earlier cities of Tadmor through excavations. Trenches excavated by Robert du Mesnil du Buisson revealed Bronze Age layers and the so-called caliciform pottery represented by characteristic

Fig. 3.7 *The Harbaqa dam photographed by Antoine Poidebard from the air in the 1920s–1930s.*
Source: Poidebard 1934

Fig. 3.8 *A Roman well at Qseybe in the Palmyrene desert-steppe area.*
Photo: Minna Lönnqvist 2006, SYGIS

Fig. 3.9 *A qanat in the Palmyrene desert-steppe area at Dedjan near the Roman site of Qebaqeb*. Photo: Kenneth Lönnqvist 2006, SYGIS

Figs. 3.10, 3.11 *Pumping water in the desert of Palmyra in 1929*. Library of Congress, Matson Collection

band-decorated goblets dating to the Early Bronze Age IV period of the late 3rd millennium BC (Fig. 3.12). The great kingdom of Ebla to the northwest has provided similar pottery, as has Qatna (see Fig. 3.13).[11] This was a special time of pastoralist movements, their cultures, and sedentarisation in Syria-Palestine.[12] Later excavations by Michel al-Maqdissi have traced more Bronze Age layers in the precincts of the temple.[13]

The archaeological evidence is concrete and confirms that there was indeed a site called Tadmor. Cuneiform inscriptions on clay tablets from the archives found at Mari, a city near the Middle Euphrates, refer to Tadmor during the Amorite rule of the kingdom (Figs. 3.13, 3.14). The Sutean nomads especially dwelt in the desert between Tadmor/Palmyra and the Euphrates as we shall further discuss in due course. The tablets from Kültepe, known as the ancient trade centre of Kanesh in Cappadocia in Anatolia, also mention Tadmor.[14] The city of Kanesh was a trading colony that included Old Assyrian merchants, such as Amorites, who were involved in the trade with their donkey caravans.[15] We shall further deal with these groups of people and individual persons among them in Ch. 4.

From the Iron Age (1200–550 BC), there are a few historical and Biblical references that have been connected with Tadmor. In Assyrian sources, Tadmor is associated with the

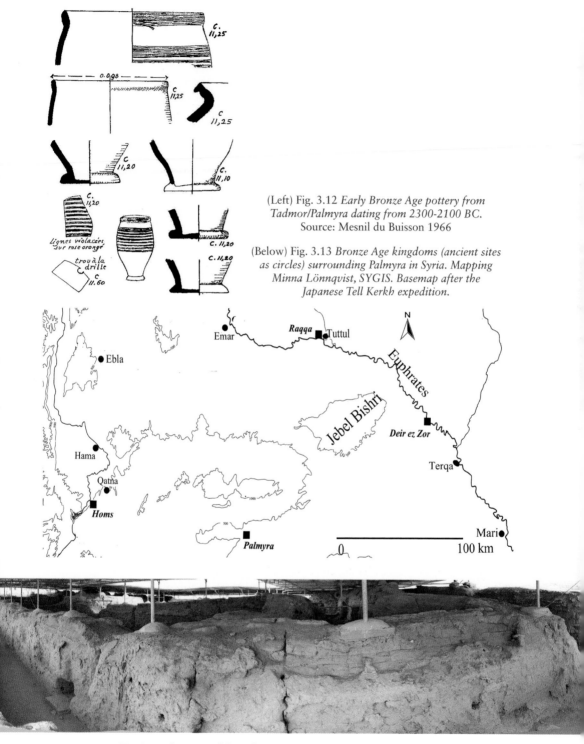

(Left) Fig. 3.12 *Early Bronze Age pottery from Tadmor/Palmyra dating from 2300-2100 BC.* Source: Mesnil du Buisson 1966

(Below) Fig. 3.13 *Bronze Age kingdoms (ancient sites as circles) surrounding Palmyra in Syria. Mapping Minna Lönnqvist, SYGIS. Basemap after the Japanese Tell Kerkh expedition.*

Fig. 3.14 *The ruins of the palace at Mari, a Bronze Age kingdom, on the Euphrates, where thousands of cuneiform tablets were found, also mentioning Tadmor in the 18th century BC.* Photo: Gabriele Fangi 2010

land and kingdom of Amurru, the areas and sites ruled by the Amorites in the Bronze Age. The first references to Arameans following the Amorites in history are associated with Tiglat-Pilesar I's campaigns in the 12th century BC. He fought battles in the district of Jebel Bishri, and was able to spread his influence as far as Tadmor. In the Assyrian sources, Arameans and Sutu appear as the people of the area.[16] The Bible states that kings David and Solomon reached the Euphrates, which was claimed as the border zone of their kingdom.[17] King Solomon has even been connected to the foundation of Tadmor.[18] Historian Josephus from the Roman period also seemed to believe in the extension of the Solomonic kingdom:

> He (Solomon) also advanced into the desert of upper Syria and having taken possession of it founded there a very great city at a distance of two days' journey from the Euphrates, while from the great Babylon the distance was a journey of six days. Now the reason for founding a city so far from the inhabited parts of Syria was that further down there was no water anywhere in the land and surrounded it with very strong walls, the named it Thadamora, as it is still called by the Syrians, while the Greeks call it Palmyra.[19]

Doubtful views concerning the identification of Tadmur in the Bible with Palmyra/Tadmor have appeared as that Tadmur has been seen as referring to a place with a similar name in Galilee. However, in the same instance there is a mention of Hamath in Syria, and Tadmor is mentioned to have been situated in a desert. Anyhow, not much archaeological evidence has been traced for the Iron Age in Palmyra so far. The Assyrian and Biblical textual references have been supplemented with a legend that the Babylonian Nebuchadnessar also visited the place.[20] The project SYGIS found Aramean pottery and an inscription at Tell Kharita on the border zone of the territory on the Euphrates.[21] Taxation of Arameans by the United Monarchy[22] in some periods may have taken place. In any case, Palmyra as a settlement began much earlier, as we have seen and dates at least from the end of the Early Bronze Age (3300–2000 BC) and was already a caravan station in the Middle Bronze Age (2000–1550 BC).

## 3.3 The Hellenistic town

It has been known for some time that Palmyra also existed in the Hellenistic period. Robert du Mesnil du Buisson, who excavated in the area of the Temple of Bel, thought that the prosperity of Palmyra started in the Hellenistic period. Pieces of Rhodian pottery imports dating from the 2nd century BC were unearthed, and the foundations of an early Temple of Bel were laid and the beginning of the cult was placed around 300 BC.[23]

The existence of this Hellenistic city had been traced in literary sources. The city already existed in the Hellenistic east which was influenced by Greek culture after the conquest of Alexander the Great. Modern technology has provided non-invasive remote sensing methods by geophysical means to trace Hellenistic structures beneath the visible Roman city of Palmyra. The remains were thought to lie outside the Roman city in a southern *wadi* that was covered by sand, and no ruins were recognisable on the ground.

Andreas Schmidt-Colinet successfully used geophysical methods in Palmyra in 1997–1998, revealing the Hellenistic city beneath the level of the Roman ruins that are visible today (see the black and white area in the southwestern part of the satellite map in Fig. 3.15). The geophysical methods (caesium magnetometry) were combined in prospecting with photogrammetric ground modeling covering an area of 2 ha. A smaller area (100 m²) was studied with resistivity. Buildings and streets were revealed.[24]

Fig. 3.15 *A satellite map of Roman Palmyra showing a radar image of the underground Hellenistic Palmyra in black and white. Sites by numbers: 1) Temple of Bel; 2) Triumphal Arch; 3) Temple of Nabu/ Nebo; 4) Bath of Diocletian; 5) Theatre; 6) Agora; 7) Senate; 8) Tetrapylon; 9) Temple of Baalshamin; 11) Basilica; 12) Houses; 13) Funerary Temple; 14) Temple of Allat; 15) Diocletian Camp; 16) Tomb of Ailami; 17) Oval Plaza; 18) Efqa Spring; 19) Monumental Column; 20) Castle.* Courtesy: GORS

Fig. 3.16 *Panorama of the Grand Colonnade in Palmyra from the south.* Photo: Gabriele Fangi 2010

Additionally, excavations were carried out in a small plot to bring actual dating evidence that confirmed the beginning of the Hellenistic occupation at least from the 3rd century BC. A caravan building (*khan*) or the residence of a caravan leader was unearthed. Architectural remains showing rich decoration such as *stucco* and frescoes (elsewhere in this way) were revealed. The finds demonstrate long-distance trade with Spain and China even then. Amphorae and fine wares, such as Megarian relief bowls and *terra sigillata*, indicate luxurious living. Rhodian, Gazan, and Egyptian wines were drunk, and olive oil and *garum* (salted fish sauce) were imported from Africa. Cooking vessels included examples from Antioch and Tarsus as well as Parthian pottery.[25]

Apart from the visible ruins on the ground, several Roman structures have been excavated in Palmyra revealing sanctuaries, public buildings, and private houses. Beneath the ground level beside the funerary structures on the surface the so-called *hypogeum* tombs have been discovered under the ground – they form subterranean *necropoleis*, cities for the dead (see Ch. 9).

## 3.4 The layout of Roman Palmyra

The city of Palmyra was surrounded by a fortification system that dates from the Hellenistic and Roman periods (see Figs. 3.15, 3.17–3.19). The walls were boundaries defining the city and its area in space. In this oasis city, it also was the boundary of the urban and settled areas with architectural planning surrounded by mobile pastoral and nomadic worlds that in turn consisted of camps and often semi-sedentary villages. Palmyra is characterised by the amalgamation of these two worlds.

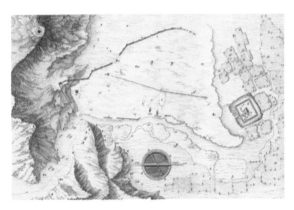

Fig. 3.17 *The map of Palmyra drawn by Giovanni Battista Borra and published in Robert Wood's book of 1753 clearly shows the main markers in the layout and built landscape of Palmyra in between the mountains and agricultural fields: the Temple of Bel, the ancient walls, the Grand Colonnade, and tower tombs.*

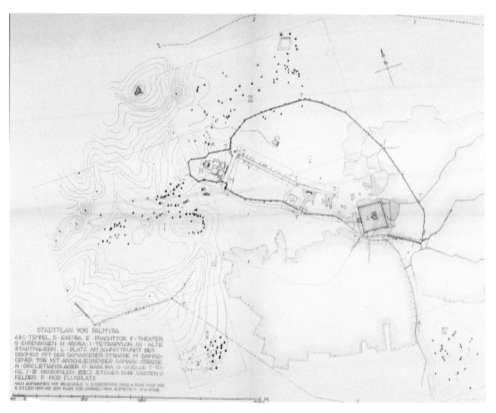

Fig. 3.18 *Map of the ruins of Roman Palmyra by Theodor Wiegand 1932.*

The ancient ramparts of Palmyra are imposing defensive features that comprise various elements. Running from north to south there is a section that rises 18 metres in height and is about 1 metre in width that was constructed of plaster or gypseous cement. This rampart has been dated to the Hellenistic period, i.e. 1st century BC.[26] The rampart resembles the innovation called *glaçis* in the older fortification systems used during the Bronze Age in the Levant.[27] As far as the walls of Palmyra are concerned, historical references to their construction exist, but matching particular parts of the walls to the references has been difficult. The major fortification system with walls, towers, and gates is now dated to the time of Emperor Diocletian and known as the Walls of Diocletian, but it has been maintained by some that at least parts of these walls date from Queen Zenobia's time and therefore the wall was called Zenobia's Wall.[28] However, the latter attribution seems to have been abandoned after the excavations of the wall which have provided material that dates the construction to the time of Diocletian.[29] There are both square and U-shaped towers in the walls of Palmyra; the square towers were built first and the U-shaped ones were later added. Moats have also been identified in the defensive system.[30]

But beside defensive purposes, walls had commercial and hydraulic significance, and were even partly used as funerary sites.[31] Some wall structures near the Diocletian Camp and the Damascus Gate (Fig. 3.18) in the south-western part of the city are called 'the

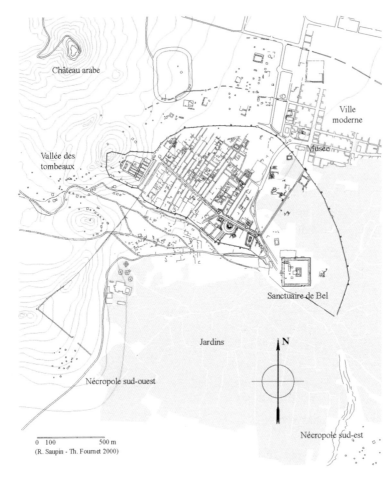

Château arabe

Vallée des
tombeaux

Ville
moderne

Musée

Sanctuaire de Bel

Jardins

N

Nécropole sud-ouest

Nécropole sud-est

0  100          500 m
(R. Saupin - Th. Fournet 2000)

(Left) Fig. 3.19
*Map of Ancient
Palmyra by
R. Saupin and Th.
Fournet 2000.*

(Below) Fig.
3.20 *The walls
of Justinian,
photographed in
1920.* Library of
Congress, Matson
Collection

Customs Wall' and have been seen as a replacement for ramparts.[32] As we have already discussed, this indeed was the district where caravans and traders entered the city from the western road through the Valley of the Tombs. As well as single tower tombs aligned along the route in the valley, tower tombs were incorporated in the structures of walls. This is an interesting feature that is also present in Hatra, where tower tombs appear in city walls.[33] Their function apparently was manifold; besides serving as a funerary place, they could have been used as watch towers for military/strategic purposes and/or star gazing/astral worship.

Palmyra had several gates; the Damascus Gate leading to the Valley of Tombs is associated with an oval plaza that is approached from a transversal colonnade leading from the Grand Colonnade to the west. The Damascus Gate is the most important gate but the strongest was apparently the Theatre Gate at the end of another transversal colonnade, thus also known as the Theatre Colonnade.[34] The Gate of Dura is in the northeast and dates from the 4th century AD. The walls, ramparts and *valla* (singular *vallum*) were restored by Emperor Justinian in the Byzantine era (Fig. 3.20).[35]

Inside the walls, the layout of Roman Palmyra has visible signs of the Hellenistic East, with colonnaded streets (Figs. 3.15, 3.17–3.19), similar to those found in Apamea (Fig. 3.22), Ephesos (Fig. 3.23), Jerash (Fig. 3.24), and Petra (Fig. 3.25). Agoras or forums, temples, theatres, baths, and *gymnasia* are typical buildings in a Greco-Roman city. Like a Greek *polis* Palmyra had a council (*boulē*), and later also a senate. The Grand Colonnade (Figs. 3.15, 3.16, 3.21; see also Ch. 7) is the main thoroughfare of the city where the majority of the known public buildings and spaces existed (see also Ch. 8). Apart from the Hippodamian grid plan comparable to the fine example in Hellenistic Dura Europos (see Ch. 5), Palmyra of the Roman period has some features from the plan of the Roman *castrum*, a military camp, that usually incorporated two main streets: the *Cardo Maximus* leading in the north–south direction and the *Decumanus Maximus* in the east–west direction. They cross at right angles in the *Tetrapylon,* a gateway structure formed by four columns.[36] Although in Palmyra the Grand Colonnade leads from the Triumphal Arch roughly from the southeast to the northwest to an imposing *Tetrapylon* (see also Ch. 7), the real *Cardo* and *Decumanus* are missing. Such are present at Apamea or in the 2nd century Jerusalem known as Emperor Hadrian's Aelia Capitolina, where the colonnaded *Cardo* (Fig. 3.26) is still visible and also appears in the famous Madaba map from Jordan. Smaller transversal streets branch from the Grand Colonnade in Palmyra. One, known as the Theatre Colonnade, leads to the west and makes a turn around the theatre offering an access to the senate.[37] Further on, a diagonally oriented transversal street begins from the *Tetrapylon* and runs towards the southwest.

The traditional layout of a military camp is, however, to be clearly found in the Camp of Diocletian (Fig. 3.15; for more details, see Ch. 10), a small area in the western part of ancient Palmyra inside the city walls, that became an important part of the garrison city in Late Antiquity. The *Via praetoria* crossed the main street, the *Via principalis,* from east to west at another *Tetrapylon*.[38]

Mudbrick was used in some Hellenistic buildings in Palmyra, for example in the oldest tomb beneath the Temple of Baalshamin dating from the 2nd century BC.[39] Stone, however,

Fig. 3.21 *The Grand Colonnade of Palmyra.* Photo: Gabriele Fangi 2010

Fig. 3.22 *A colonnaded street in Apamea in Syria.* Photo: Gabriele Fangi 2010

Fig. 3.23 *A colonnaded street in Ephesos in Turkey.* Photo: Ahmet Denker

Fig. 3.24 *The colonnaded street in Jerash in Jordan.* Photo: Kenneth Lönnqvist 1996

Fig. 3.25 *The colonnaded street of Petra in Jordan.* Photo: Kenneth Lönnqvist 1996

Fig. 3.26 *The Cardo in Jerusalem in Israel.* Photo: Kenneth Lönnqvist 1993

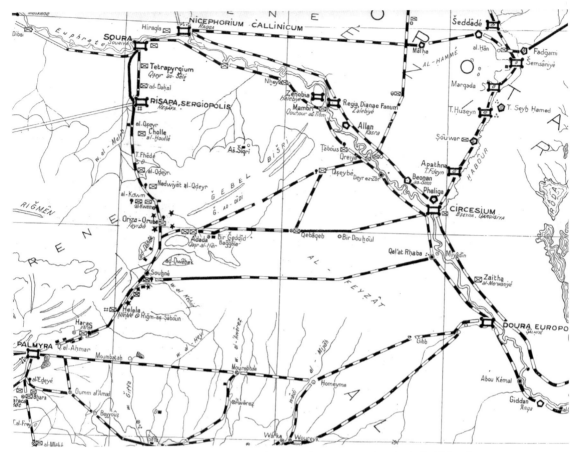

Fig. 3.27 *Sites and roads connecting Palmyra with Roman stations in A. Poidebard's map.*
Source: Poidebard 1934

was the main building material used in the ruins that are visible today. The limestone used in construction of the city, its buildings, and streets was brought from nearby, 15 km northeast of the city, where quarries have been identified. *Graffiti* on the walls of the masons' and quarry workers' habitation areas tell about the individuals that worked in quarries; elaborate systems supplied the workers with water.[40] But aside from limestone and sandstone there was a special splendour in using stone like granite transported from Aswan in Egypt, or other granite from Anatolia, as well as marble. The granite was visible in the imposing columns of the *Tetrapylon* and the portico in front of the Baths of Diocletian. Marble is found, for example, on the facing of walls and floors in the style of *opus sectile*, marble slabs forming geometric patterns in the Baths of Diocletian.[41]

## 3.5 The territory of Palmyra

The territory of Palmyra changed its status, from the Greek type of *polis* (CIS 3923, Inv. IX.8),[42] a city-state incorporating surrounding areas as the Roman *regio Palmyrena* (AD 17), to a Roman *colonia* (AD 203) later acquiring an empire of its own (AD 269/270–272). The annexation to Rome has been assumed to have started from the building project of the Temple of Bel in Palmyra in AD 17, or even earlier in AD 14.[43] However, Palmyra's culture and taste expressed its own identity[44] – a mixture of the Greco-Roman world with Persian and Egyptian influences that affected the Near East from the 4th century BC to the 3rd century AD.

Several other small villages and military sites from the Roman period are known from the desert-steppe surrounding Palmyra. In the immediate vicinity of Palmyra, Roman sites were built near watering points and oases. Ptolemy's Geography, dating from the 2nd century AD, an ancient list of sites produced as a map, mentions the following towns in the Palmyrene: Resapha, Cholle, Oriza, Putea, Adada, Palmyra, Adacha, Danaba, Goaria, Aueria, Casama, Odmana, and Atera.[45] Some of them have been identified by Alois Musil[46] and Antoine Poidebard[47] (see Fig. 3.27) on the ground, many connected to the *Strata Diocletiana*: Araq, Sukhna, Taiba, Qasr-al-Hayr ash-Sharqi, El-Kowm, and Resafa, as well as Qebaqeb and Qseybe. Several of the sites were military posts, *castri* and *castelli*.[48] John F. Matthews and Andrew M. Smith have discussed the limits of Palmyra from the epigraphical point of view, covering areas as far as Emesa and Qasr al-Hayr al-Gharbi in the west, and Smith also points to the evidence in the east.[49] Using epigraphic evidence, Leonardo Gregoratti has also recently traced the territory of the city.[50]

After Alois Musil's, Antoine Poidebard's, and Daniel Schlumberger's[51] studies on the ground, the project SYGIS mapped Roman military and village sites in the region of Jebel Bishri in the Palmyrides using remote sensing methods, such as satellite image data sources and field work applying GIS (Geographic Information Systems) in mapping in 2000–2010.[52] That was the introduction of GIS[53] to field work in the region. A Syrian–Norwegian team, with leadership that originally used to work under SYGIS, has followed in the footsteps of the project and has detected further Roman sites in the Palmyrides, using Google Earth for mapping.[54] In addition, Stefan Hauser[55] has dealt with Palmyra as a territorial power using Google Earth mapping. Neither Matthews, Smith, nor Hauser have, however, included the areas as far as the Euphrates as listed by Ptolemy in his Geography. Archaeological evidence of the territorial influence of Palmyra that we shall further discuss in Ch. 4, however, reaches over Jebel Bishri to the Euphrates, and also to the famous fortress of Zenobia (Ch. 5).

The walls, the existence of the Camp of Diocletian, and the *Strata Diocletiana* show that military interests and protection of the city were the focus in the Late Roman times in the 4th century AD (see more in Ch. 10).

## Endnotes

1    Kenyon, M. (1959) Some Observations on the Beginnings of Settlement in the Near East, in *Journal of Royal Anthropological Society*, Vol. 89, pp. 35–43, p. 40.

2    Hodder, I. (2006) *The Leopard's Tale: Revealing the Mysteries of Çatalhöyük*. London: Thames and Hudson.

3    Sanlaville, P. and Trabouli, M. (1996) Palmyre et la Steppe Syrienne, in *Les Annales Archéologiques Arabes Syriennes*, Vol. XLII, 1996, *Special Issue Documenting the Activities of the International Colloquium Palmyra and the Silk Road*, pp. 29–40.

4    Smith, S.L., Wilkinson, T.J. and Lawrence, D. (2014) Agro-pastoral Landscapes in the Zone of Uncertainty: The Middle Euphrates and the North Syrian Steppe during the 4th and 3rd millennia BC, in *Settlement*

*Dynamics and Human-Landscape Interaction in the Dry Steppes of Syria*, ed. by Morandi Bonacossi, D., Wiesbaden: Harrassowitz Verlag, pp. 151–172.

5   Du Mesnil du Buisson, R. (1966) Prèmiere campagne de fouilles a Palmyre, in *Comptes rendus des séances de l'Académie des Inscriptions et Belles-Lettres*, Vol. 110, pp. 158–190.

6   Al-As'ad, Kh. and Gawlikowski, M. (1997) *The Inscriptions in the Museum of Palmyra: A Catalogue*, Palmyra & Warsaw: The Committee for Scientific Research, p. 21.

7   Crouch, D. P. (1975) The Water System of Palmyra, in *Studia Palmyreńskie*, Vol. VI–VII, pp. 151–186; Juchniewicz, K. and Zuchowska, M. (2012) Water Supply in Palmyra, A Chronological Approach, in *The Archaeology of Water Supply*, ed. by Zuchowska, M., BAR International Series 2414, Oxford: Archaeopress, pp. 61–73.

8   Hammad, M. (2010) *Palmyre: Transformations urbaines, Développement d'une ville antique de la marge aride syrienne*, Paris: Geuthner, pp. 10–11.

9   See Musil, A. (1928) *Palmyrena: A Topographical itinerary*, American Geographical Society, Oriental Explorations and Studies No 4, New York: American Geographical Society, p. 131–; Musil, A. (1927) *The Middle Euphrates: A Topographical itinerary*. American Geographical Society. Oriental Explorations and Studies No 3. New York: American Geographical Society; see also Poidebard, A. (1934) *La trace de Rome dans le désert de Syrie: Le limes de Trajan à la conquête arabe. Recherches aériennes (1925–1932)*. BAH. Tome XVIII. Texte and Atlas. Paris: Paul Geuthner.

10  Lönnqvist, M., Törmä, M., Lönnqvist, K. and Nuñez, M. (2011) *Jebel Bishri in Focus: Remote sensing, Archaeological surveying, mapping and GIS studies of Jebel Bishri in central Syria by the Finnish project SYGIS*, BAR International Series 2230, Oxford: Archaeopress.

11  Du Mesnil du Buisson, R. (1966) Prèmiere campagne de fouilles a Palmyre, in *Comptes rendus des séances de l'Académie des Inscriptions et Belles-Lettres*, Vol. 110, pp. 158–190; see also Du Mesnil du Buisson, Le Comte (1967) Decouverte de La Plus Ancienne Palmyre, Ville Amorite de La Fin Du IIIe Millenaire, in *Archeologia* 16, pp. 50–51.

12  See, for example, Lönnqvist, M. (2008) Were Nomadic Amorites on the Move? in *Proceedings of the 4th International Congress of the Archaeology of the Near East, 29 March – 3 April 2004, Freie Universität Berlin*, Vol. 2, Social and Cultural Transformation: The Archaeology of Transitional Periods and Dark Ages, Excavation reports, ed. by Kühne, H., Czichon, R. and Kreppner, F. J., Wiesbaden: Harrassowitz Verlag, pp. 195–215.

13  Al-Maqdissi, M. (2000) Note Sur les Sondages Réalisés par Robert Du Mesnil Du Buisson dans la Cour Du Sanctuaire de Bêl À Palmyre, in *Syria*, Vol. 77, pp. 137–154.

14  See Klengel, H. (1996) Palmyra and International Trade in the Bronze Age: the Historical Background, in *Les Annales Archéologiques Arabes Syriennes*, Vol. XLII, 1996, *Special Issue Documenting the Activities of the International Colloquium Palmyra and the Silk Road*, pp. 159–163, especially p. 159.

15  See Lewy, J. (1952) Studies in the Historic Geography of the Ancient Near East: Old Assyrian Caravan Roads in the Valleys of Habur and the Euphrates and in Northern Syria, in *Orientalia*, Vol. XXI, pp. 265–292.

16  Brinkman, J.A. (1968) *A Political History of Post-Kassite Babylonia, 1158–722 BC*, Analecta Orientalia 43, Commentationes scientificae de Rebus Orientis Antiquitis, Roma: Pontificium Institutum Biblicum, pp. 267–285; see also Lawson, K. Younger Jr. (2016) *A Political History of Arameans: From their Origins to the End of their Polities*, Atlanta: SBL Press, pp. 36–37.

17  II Samuel 8: 3-6; I Kings 4:21; II Chronicles 9: 26.

18  I Kings 9:18; II Chronicles 8: 4.

19  Josephus, *Jewish Antiquities*, Book VIII, 150–154, the Loeb Classical library, transl. Thackeray, H. St.J. and Marcus, R., 1977, Cambridge, Mass.: Harvard University Press.

20  See du Mesnil du Buisson, R. (1966) Prèmiere campagne de fouilles a Palmyre, in *Comptes rendus des séances de l'Académie des Inscriptions et Belles-Lettres*, Vol. 110, pp. 158–190, especially p. 185.

21  Lönnqvist, M., Törmä, M., Lönnqvist, K. and Nuñez, M. (2011) *Jebel Bishri in Focus: Remote sensing, archaeological surveying, mapping and GIS studies of Jebel Bishri in central Syria by the Finnish project SYGIS*. British Archaeological Reports International Series 2230, Oxford: Archaeopress, pp. 233–237.

22  See II Samuel 8: 3–6.

23  Du Mesnil du Buisson, R: (1966) Prèmiere campagne de fouilles a Palmyre, in *Comptes rendus des séances de l'Académie des Inscriptions et Belles-Lettres*, Vol. 110, pp. 158–190, especially p. 184.

24  Schmidt-Colinet, A. and Al-As'ad, Kh. (2002) Archaeological News from Hellenistic Palmyra, in *Parthica*, Vol. 4, pp. 157–166; Schmidt-Colinet, A., Al-As'ad, K. and Al-As'ad, W. (2016) Palmyra, 30 Years of Syro-German/Austrian Archaeological Research (Homs), in *A History of Syria in One Hundred Sites*, ed. by Kanjou, Y. and Tsuneki, A., Oxford: Archaeopress, pp 339–348, especially pp. 345–347. Some technical details are mentioned

in an abstract submitted by Schmidt-Colinet, A. in the CIPA Workshop for Saving the Heritage of Syria in the 10th ICAANE in Vienna 2016, organized by Silver, M. and Doneus, M.; see the abstract booklet: https://www.oeaw.ac.at/fileadmin/Institute/OREA/pdf/events/ICAANE_2016_Abstract_Booklet.pdf; see also Linck, R. (2016) Geophysical Prospection by Ground- and Space-based Methods of the Ancient Town of Palmyra (Syria), in *Palmyrena: City, Hinterland and Trade between Orient and Occident, Proceedings of the Conference Held in Athens, December 1–3, 2012*, ed. by Meyer, J.C., Seland, E.H. and Anfinset, N., Oxford: Archaeopress, pp. 77–85.

25  Schmidt-Colinet, A. and Al-As'ad, Kh. (2002) Archaeological News from Hellenistic Palmyra, in *Parthica*, Vol. 4, pp. 157–166; Schmidt-Colinet, A., Al-As'ad, Kh. and Al-As'ad, W. (2016) Palmyra, 30 Years of Syro-German/Austrian Archaeological Research (Homs), in *A History of Syria in One Hundred Sites*, ed. by Kanjou, Y. and Tsuneki, A., Oxford: Archaeopress, pp. 339–348, especially pp. 345–347.

26  Crouch, D. P. (1975) The Ramparts of Palmyra, in *Studia Palmyreńskie*, Vol. VI–VII, pp. 6–44; Gawlikowski, M. (1975) Remarks on the Ramparts of Palmyra, in *Studia Palmyreńskie*, Vol. VI–VII, pp. 45–46.

27  See, for example, Kaplan, J, (1975) Further Aspects of the Middle Bronze Age II Fortifications in Palestine, in *Zeitschrift des Deutschen Palästina-Vereins*, Vol. XCI, pp. 1–17.

28  See Bounni, A. and Al As'ad, Kh. (1997) *Palmyra: History, Monuments & Museum*, Damascus, pp. 90–91.

29  Juchniewicz, K., As'ad, Kh. and Al Hariri, K. (2010) The Defense Wall in Palmyra after Recent Syrian Excavations, in *Studia Palmyreńskie*, Vol. XI, pp. 55–73.

30  Juchniewicz, K., As'ad, Kh. and Al Hariri, K. (2010) The Defense Wall in Palmyra after Recent Syrian Excavations, in *Studia Palmyreńskie*, Vol. XI, pp. 55–73.

31  Hammad, M. (2013) The morphology of the environs of Palmyra: ground relief, environment, roads, in *Studia Palmyreńskie*, Vol. XII, *Fifty Years of Polish Excavations in Palmyra 1959–2009, International Conference, Warsaw, 6–8 December 2010*, pp. 129–148, see walls pp. 133–136.

32  See Bounni, A. and Al As'ad, Kh. (1997) *Palmyra: History, Monuments & Museum*, Damascus, pp. 90–91.

33  Juchniewicz, K., As'ad, Kh. and Al Hariri, K. (2010) The Defense Wall in Palmyra after Recent Syrian Excavations, in *Studia Palmyreńskie*, Vol. XI, pp. 55–73, especially p. 56; see also Silver, M, Törmä, M., Silver , K., Okkonen, J., Nuñez, M. (2015) The Possible Use of Ancient Tower Tombs as Watchtowers in Syro-Mesopotamia, in *ISPRS Annals* (II-5/W3), ed. by Yen, Y-.N., Weng, K.-H and Cheng, H.-M., pp. 287–293.

34  Juchniewicz, K., As'ad, Kh. and Al Hariri, K. (2010) The Defense Wall in Palmyra after Recent Syrian Excavations, in *Studia Palmyreńskie*, Vol. XI, pp. 55–73, especially p. 58.

35  Hammad, M. (2013) The morphology of the environs of Palmyra: ground relief, environment, roads, in *Studia Palmyreńskie*, Vol. XII, *Fifty Years of Polish Excavations in Palmyra 1959–2009, International Conference, Warsaw, 6–8 December 2010*, pp. 129–148; see walls pp. 133–136.

36  See, for example, Stambaugh, J.E. (1988) *The Ancient Roman City*, Baltimore and London: Johns Hopkins University Press, p. 270, 283; see also Ball, W. (2016) *Rome in the East: The Transformation of an Empire*, 2nd edition, London, New York: Routledge, p. 327.

37  See, for example, Browning, I. (1979) *Palmyra*, London: Chatto & Windus, pp. 81–87; Bounni, A. and Al-As'ad, Kh. (1988) *Palmyra: History, Monuments and Museums*, Damascus, pp. 74–76.

38  See, for example, Browning, I. (1979) *Palmyra*, London: Chatto & Windus, p. 185.

39  Baranski, M. (1996) Development of the Building Techniques in Palmyra, in *Les Annales Archéologiques Arabes Syriennes*, Vol. XLII, 1996, *Special Issue Documenting the Activities of the International Colloquium Palmyra and the Silk Road*, pp. 379–384.

40  Schmidt-Colinet, A. (1995) The Quarries of Palmyra, in *ARAM*, Vol. 7, pp. 53–58; Schmidt-Colinet, A., Al-As'ad, Kh. and Al-As'ad, W. (2016) Palmyra, 30 Years of Syro-German/Austrian Archaeological Research (Homs), in *A History of Syria in One Hundred Sites*, ed. by Kanjou, Y. and Tsuneki, A., Oxford: Archaeopress, pp. 339–348, especially p. 341.

41  Apart from the Baths of Diocletian, marble appears in various decorations in Palmyra and a local marble quarry has also been discovered: see Dodge, H. (1988) Palmyra and the Roman Marble Trade: Evidence from the Baths of Diocletian, in *Levant*, Vol. XX, pp. 215–230; see also Wielgosz, D. (2013) Coepimus et lapide pingere: marble decoration from the the so-called "Baths of Diocletian" at Palmyra, in *Studia Palmyreńskie*, Vol. XII, pp. 319–332.

42  Yon, J.-B. (2002) *Les Notables de Palmyre*, BAH 163, Beyrouth: Ifpo, p. 9.

43  See Smith, A.M. II (2013) *Roman Palmyra: Identity, Community, and State Formation*, Oxford: Oxford University Press, p. 2; see also Stoneman, R. (1995) *Palmyra and its Empire: Zenobia's Revolt against Rome*, Ann Arbor: The University of Michigan Press, p. 27, pp. 201–202; see also the terminology in Sartre, M. (1996) Palmyre, cité grecque, in *Les Annales Archéologiques Arabes Syriennes*, Vol. XLII, 1996, *Special Issue Documenting the Activities of the International Colloquium Palmyra and the Silk Road*, pp. 385–405. In connection of Palmyra the term *polis* is used in various occasions from the 1st century AD continuing to be used in the 3rd century AD.

44  See Anadol, S. (2008) Palmyra – Identity Expressed through Architecture and Art, in *Jebel Bishri in Context, Introduction to the Archaeological Studies and the Neighbourhood of Jebel Bishri in Central Syria, Proceedings of a Nordic Research Training Seminar in Syria, May 2004*, ed. by Lönnqvist, M., BAR International Series 1817, Oxford: Archaeopress, pp. 59–72; see Smith, A.M. II (2013) *Roman Palmyra: Identity, Community, and State Formation*, Oxford: Oxford University Press.

45  Claudius Ptolemy, *The Geography*, transl. and ed. by Stevenson, E.L., New York: Dover Publications, Locations in Syria, Book V, Ch. XIV, p. 127.

46  Musil, A. (1928) *Palmyrena: A Topographical itinerary*. American Geographical Society. Oriental Explorations and Studies No. 4. New York: American Geographical Society; Musil, A. (1927) *The Middle Euphrates: A Topographical itinerary*. American Geographical Society. Oriental Explorations and Studies No 3. New York: American Geographical Society.

47  Poidebard, A. (1934*) La trace de Rome dans le désert de Syrie: Le limes de Trajan à la conquête arabe. Recherches aériennes (1925–1932)*. BAH. Tome XVIII. Texte and Atlas. Paris: Paul Geuthner.

48  Poidebard, A. (1934*) La trace de Rome dans le désert de Syrie: Le limes de Trajan à la conquête arabe. Recherches aériennes (1925–1932)*, BAH. Tome XVIII. Texte and Atlas. Paris: Paul Geuthner.

49  Matthews, J.F. (1984) The Tax Law of Palmyra: Evidence of Economic History from a City in the Roman East, in *Journal of Roman Studies*, Vol. 74, pp. 157–180; Smith, A.M. II (2013) *Roman Palmyra: Identity, Community, and State Formation*, Oxford: Oxford University Press, p. 2.

50  Gregoratti, L. (2015) *Palmyra, City and Territory through the Epigraphic Sources, in Broadening the Horizons: A Conference of young researchers working in the Ancient Near East, Central Asia, University of Toronto, October 2011*, ed. by Affanni, G., Baccarin, C., Cordera, L., Di Michele, A. and Kavagnin, K., BAR International Series 2698, Oxford: Archaeopress, pp. 55–59.

51  Schlumberger, D. (1951) *La Palmyrène du Nord-Ouest, suivie du Recueil des inscriptions sémitiques de cette région* par H. Ingholt et J. Starcky, avec une contribution de G. Ryckmans. Paris: Geuthner.

52  Lönnqvist, M., Törmä, M., Lönnqvist, K. and Nuñez, M. (2011) *Jebel Bishri in Focus: Remote sensing, archaeological surveying, mapping and GIS studies of Jebel Bishri in central Syria by the Finnish project SYGIS*. British Archaeological Reports International Series 2230, Oxford: Archaeopress.

53  See the applications of Geographic Information Systems (GIS) in archaeological documentation in Lönnqvist, M. and Stefanakis, E. (2010) ) GIScience in Archaeology: Ancient Human Traces in Automated Space, in *The Manual of Geographic Information Systems*, ed. by Madden, M., Bethesda, Maryland: American Society of Photogrammetry and Remote Sensing, pp. 1221–1259.

54  Anfinset, N. and Meyer, J. C. (2010) The hinterland of Palmyra, in *Antiquity*, No. 324, Project Gallery articles.

55  Hauser, S.R. (2012) Wasser als Ressource: Palmyra als Territorialmacht, in *Wasserwirtschafliche Innovationen im archäologischen Kontext: Von den prähistorischen Anfängen bis zu den Metropolen in Antiken*, Forschungscluster 2, Innovationen: technisch, sozial, Herausgegeben von Klimscha, F., Eichmann, R., Schuler, C. and Fahlbusch, H., Rahden/Westfalen: Deutsches Archäologisches Institut, pp. 211–224.

# 4

# TRIBES, FAMILIES, AND INDIVIDUALS IN INSCRIPTIONS AND IMAGES

## 4.1 Nomadism, tribes, clans, and families

As Palmyra/Tadmor is an oasis in the Syrian Desert we have seen in previous chapters how the environment has affected the livelihood of the people in the neighbourhood. Mobility has been an ongoing factor in the region throughout its history, first in the lives of hunter-gatherers and then with pastoral nomads, who have been traversing the landscape of the desert and steppe. Pastoral nomadism has been based on tribal social organisation in the region for thousands of years. Families and clans are held in esteem (Fig. 4.1). Tribalism also concerns the rule of territories and grazing grounds.

Fig. 4.1 *The funerary couch of Bolbarak family displaying a banquet scene dated to AD 240.*
Photo: Kenneth Lönnqvist 2004, SYGIS

Fig. 4.2 *Pastoral nomads grazing in the area of the Palmyrides in spring time.*
Photo: Minna Lönnqvist 2003, SYGIS

As previously discussed, Tadmor (Palmyra) as a caravan city had roots long before becoming a station in the Silk Road. Trading is natural for the economy of pastoral nomads (Fig. 4.2) as nomads often need products of the neighbouring settled population,[1] while in turn pastoral products are used by town and city dwellers. As already mentioned (Ch. 2), because of mobility and trading nomads also have become professionals in transporting commodities.

Monuments visible in the area have provided information about the kind of people that lived in Palmyra during the Greco-Roman period. The numerous inscriptions and funerary portraits are valuable sources to connect us with the ancient people, the society of the city, and its life. As earlier indicated, there are plenty of inscriptions found inside and around the city. Especially in the Grand Colonnade, honorific inscriptions referring to powerful people were incised on columns that in antiquity carried statues on brackets (Fig. 4.3). But inscriptions were also cut on lintels (Figs. 4.5, 4.6), walls, and altars as well as funerary monuments and items including sarcophagi, slabs, and tombstones.[2] Inscriptions mainly appear in three languages: Palmyrene Aramaic, Greek, and Latin (Figs. 4.4–4.7), but Arabic has also been used, for example in nearby Bedouin camps[3] (Fig. 4.8). Detailed studies of private houses and tombs have brought the ancient families and clans of Palmyra even closer, providing a picture of the past society and its lifestyle. Such detailed information of the whole community in an ancient city is a treasure trove.

As discussed in the previous chapter, the remains that are mostly encountered on the surface of Palmyra today culturally have a more ancient heritage than has been generally acknowledged in works concentrating on Greco-Roman Palmyra. We can dig deeper into the past to understand the cultural amalgamation that has taken place in this oasis city. The questions of the invisibility of nomads in history and archaeology have been discussed in recent decades, and, as already seen, new methods to trace material culture of pastoral nomads were developed in the 1980s and 1990s.[4] It is important to visualise the past that has been neglected and left invisible. The nomadic culture of Palmyra flourished and many other caravan cities that we have touched on in the present work; some on the Silk Road, others on the Incense and Spice Road, developing from local backgrounds and traditions. Through a longer and wider perspective, we can find how vibrant the culture and the impact of the surrounding and intermingling nomads have been and what viable impact they have brought to the cultures of the Near East and sites like Palmyra.[5]

Fig. 4.3 *A statue standing on a column bracket in Palmyra.* Photo: Gabriele Fangi 2010

Fig. 4.4 *A trilingual inscription inscribed in Latin, Greek and Aramaic that was kept in the garden of the Palmyra Museum. The inscription mentions Statilius and Alcimus, the latter also appears as a tax collector in the famous Tax Law of Palmyra.* Photo: Kenneth Lönnqvist 2004, SYGIS, Courtesy of the Palmyra Museum

Fig. 4.5 *An Aramaic inscription in the door lintel of a hypogeum tomb in Palmyra.* Photo: Silvana Fangi 2010

Fig. 4.6 *A Greek inscription in Palmyra.* Photo: Silvana Fangi 2010

Fig. 4.7 *A Latin inscription from the Homs–Palmyra track dating from the 3rd century AD and photographed in 1929.* Library of Congress, Matson Collection

Fig. 4.8 *Arabic Bedouin inscriptions in the Palmyrene desert-steppe at the site of a camp along caravan roads.* Photo: Nils Anfinset 2004, SYGIS

Therefore, we wish in the present work to take a broader view and emphasise the importance of the continuity in the region of Palmyra that has already partly been presented in the previous chapters. The tribal roots and Semitic heritage are thousands of years old at the site, without which the Roman Palmyra could not have its own peculiar character. It needs to be mentioned that the society of Palmyra was still focused on clans and families before ISIS took over. Evidence for continuity covers the whole of historical time. The As'ad family, for example, was a powerful clan and family in Palmyra, and the directorship of the Palmyra Museum was inherited by Waleed from his father Khaled.

## 4.2. The Amorites, Suteans, and Arameans

Ancient cuneiform texts from the Mari palace archives (ARM) on the Euphrates have revealed the role of pastoral nomads in the region of Palmyrena, the Middle Euphrates, and the Jezira – the area between the Twin Rivers. The Amorite tribes were called Martu in Sumerian and Amurru in Akkadian – meaning 'westerners' – personified in their god Martu/Amurru (see Fig. 4.9). The kings of Mari were derived from Amorite tribes during the Amorite dynasty. They formed the major population of the region, and the urban kingdoms enclosed pastoral nomads in their domain in the 18th century BC. The pastoralists were supplying raw materials to the kingdoms. Earlier the tribes had been involved in bringing down the Sumerian civilisation in Southern Mesopotamia c. 2000 BC.[6] Nomadism and tribalism in the region not only included trading activities but pastoralists also served as mercenaries and escorts for more settled societies in the Middle Bronze Age (2000–1550 BC).[7]

An Old Assyrian document from the early 2nd millennium BC and the Middle Bronze Age refers to a man from Tadmor bearing a Semitic name Puzur-Ishtar. He is the first Tadmorean known by name, and his seal has been identified on an Old Assyrian clay envelope. There is also a document that testifies that a Tadmorean had delivered one mina of silver (nearly 0.5 kg).[8] So, these texts refer to the previously indicated fact that Tadmoreans included Semites and were connected in commercial transactions with Kültepe-Kanesh, an Assyrian trading post in Anatolia, in the early 2nd millennium BC. An inhabitant from Tadmor named Ulluri is also mentioned in the Mari texts (ARM XXII, 15 III, 11).[9]

The Suteans, a tribal group of people, are specifically known to have populated the desert area from Tadmor to the Euphrates in the Middle Bronze Age. Several sites of the Suteans depend on the existence of wells. The Suteans were involved in sheep breeding, and their taxes were paid to the Mari kingdom on the Euphrates in the form of sheep, donkeys, and textiles. The chief of the Suteans, called Šamiš, is mentioned in the context of tax payments.[10] A man known as Mut-Bisir, a personal name including the toponym of Jebel Bishri, appears providing truffles and ostrich eggs to the king from the region.[11] Desert truffles are a great delicacy; they are

Fig. 4.9 *The God Amurru, the personification of the Amorite people, as inscribed in a seal.* Redrawn by Minna Lönnqvist from Amiet 1980

among the expensive fungi of the region and were worthy of the king's table. They have been collected by Bedouins in the Jebel Bishri region until recent years.[12] In addition, among the pastoralists the Suteans appear to be mercenaries who escorted caravans, according to the Mari sources, and in this case two chiefs Gaidanum and Ili-epuh are mentioned related to the activities with King Shamshi-Adad.[13] However, the Suteans also attacked sites near Palmyra; they were involved in capturing slaves and participated in the slave trade. The Amorites and Suteans lived side by side, but there is no consensus as to whether Suteans actually belonged to the Amorite tribes living in the Syrian Desert and the Middle Euphrates. They particularly appear in the region of Jebel Bishri that became a nomadic outpost and was a deified mountain with numerous animal pens and cairns/*tumulus* tombs.[14]

The Amorites and Suteans were followed by the Arameans, who also lived in the region of the Palmyrides and the Syrian Desert, in the Late Bronze Age (1550–1200 BC) and the Iron Age (1200–550 BC). Several scholars have suggested that Arameans, in fact, are descendants of the Amorites and/or Suteans. Their major settlement areas and West Semitic dialects can be seen overlapping in the region. When the Amorites and Suteans disappear from history, the Arameans emerge.[15] We have already mentioned the campaigns of Tiglat-pileser I in the region. At that time Palmyra was considered to be part of the kingdom of Amurru[16] which originally consisted of Amorites in Central Syria in the Middle Bronze Age but later in the Late Bronze Age reached the Mediterranean coast.[17] The Arameans for their part built important kingdoms in the Euphrates region during the Assyrian period, and their pastoral groups were in conflict with the Mesopotamian rulers like the Amorites had been before.[18] There are indications that later on Palmyra was sacked by the Babylonians. The Persian period was the great revival and the internationalisation of the Aramean culture, when Aramaic language became *lingua franca* for the whole East.[19]

## 4.3 People of the Greco-Roman period:
## Palmyrene Arameans, Jews, and Arabs

There have been estimations of the population size of Roman Syria in the 1st century AD. The suggestions have varied from 2 million to 5 million people, and great cities like Apamea could have had 100–125,000 inhabitants, taking into account the urban and surrounding rural areas.[20] How about Palmyra? It could have been much smaller at the time but it probably grew in the 2nd and 3rd centuries AD. Kevin Butcher has visually analysed the extent of the city in the Roman times and compared it to other contemporary centres in Syria-Palestine. Palmyra appears to have been smaller than Antioch but comparable in size with Damascus.[21] In recent decades the population of the city has only been around 30,000 people.[22] Judging from its character as a mere village in the 18th to early 20th century the number of inhabitants was probably much less then, but we need to take into account the pastoral nomads of the surrounding desert belonging to its influence and in turn influencing a village or a small town.

The ancient community leaders and members of Palmyra are known in thousands by their tribal names, family names, and private names. The study of names and families is called prosopography. Sir Ronald Symes was a leading expert on the Roman prosopography of Latin names. Beside the Aramaic names, there are Greek, Latin, and sometimes Arabic or Iranian names in Palmyra. Arabic words also appear in Palmyrene inscriptions. [23] As already mentioned, after the time of the Amorites, Arameans started dominating the region of Palmyra, and in due course Arabs, who for the first time appear in the Assyrian sources,[24] gradually migrated to the area[25] and were there during the Greco-Roman period as Strabo explains in his Geography.[26] In the Greco-Roman Palmyra a person could have had either a Greek or a Latin name as well as an Aramaic one. Family trees can be built based on the inscriptions preserved in Palmyra, and one can study the relationship of families and their members. Families were patrilineal, meaning that the descent passed through the male members of the family in an Oriental way so that names were transferred from a father to a son. *Br* appears in names referring to a son,[27] cf. in Hebrew Bar-Nathan, Bar-Yosef, the son of Nathan, the son of Yosef as family names. On the other hand, Cynthia Finlayson holds that the society of Palmyra in the Roman times was matriarchal.[28] It is true that female members of the society had been given an imposing position, and Queen Zenobia is an example of female power achieved in the city that even formed an empire of its own in the Near East. This, on the other hand, does not mean that the patrilineal descent was not followed, and visible in the nomadic heritage of the region; even today, patrilocalism is met, meaning that a son stays in the neighbourhood of father's house after his marriage.[29]

Intermarriage with family members took place, and was especially common in priestly families. Girl could have married in their early teens, and an aunt could marry her nephew. Professions and offices of many of the Palmyrene people are also known; families formed a political entity in Palmyra in the 1st century AD.[30] Besides Aramaic and Arab populations, there are indications of Palmyrene Jews from that period in the epitaphs of Palmyra and in the Beth Shearim catacombs in Palestine. One is mentioned as having been a banker and others may have been traders, but generally Palmyra's conquests and rule in Mesopotamia in the 3rd century AD were seen as bad effects for Jews and their trade in the region.[31]

## 4.4 Galleries of ancient portraits

The family tombs in Palmyra reflect the unity and importance of such groups like tribes, clans and families in Palmyra (see Ch. 9). There was tribal interdependence in the polis.[32] A great number of individual portraits, many associated with funerary inscriptions, have been discovered in Palmyra (see Figs. 4.10, 4.12–4.23).[33] They provide a valuable source for studying the families and individuals in the great caravan city. In portraiture the status and hierarchy in the family and the society are visible.[34] Priests were especially valued (Fig. 4.20). The tribe of Komare was a powerful priestly clan in Palmyra during the Roman period, the evidence of which starts from the 1st century AD.[35] Apart from administrators, trading families had a special power in the city and its territory as well.[36]

Parents and children are represented in sculptured family portraits on sarcophagi (coffins made of hard material) and funerary slabs (Figs. 4.1, 4.11). The members of the families often appear in a hierarchical manner, and in that order children in family groups appear as very small, not in proportion. This also concerns portraits of individual women presented with a

Fig. 4.10 *Ancient funerary portraits of Palmyrenes photographed in 1867.*
Photo: Library of Congress, Maison Bonfils, Beirut, Lebanon

Fig. 4.11 *Funerary sculpture, portraits of family members illustrated under a funerary couch, photographed in Palmyra in 1920 during the French Mandate period.* Library of Congress, Matson Collection

child that is indicated in a smaller size than in life or in proportion. It seems that in those cases children appear as attributes. But there also are funerary portraits of young children, who are illustrated individually and larger (Figs. 4.12–4.16). They often carry a dove similar to children on Greek grave steles, indicating to a dove that was a token for goddess Dione.[37] Also, a bunch of grapes appear in the portraits of the children that may refer to Dionysos as the god of the wine, the underworld, and the fertility for new life in nature. The Palmyra Museum also contained sculptured heads as well as funerary relief types of steles. Unfortunately, looting of Palmyrene portraits has taken place during the civil war of Syria, and such portraits have been caught up, smuggled to Lebanon. Even the Syrian army was involved this kind of antiquities trafficking before ISIS conquered Palmyra.

Harald Ingholt's classification of the Palmyrene portraiture *Studies over Palmyrene Sculpture* (*Studier over Palmyrensk Skulptur* in the original Danish),[38] is a classic work that dates from 1928. It provides traditional and chronological guidelines for dating the styles of the Palmyrene busts and reliefs of people that mainly come from funerary contexts. Ingholt's

Figs. 4.12–4.16 *Children and youths carved on tombstones from the Palmyra Gallery of Portraits dating from the Roman Period* Photographs Kenneth Lönnqvist 2009, Courtesy of the Palmyra Museum

Figs. 4.17.–4.23 *Adults from the Palmyra Gallery of Portraits dating from the Roman Period.* Photos: Kenneth Lönnqvist 2004–2008, SYGIS, Courtesy of the Palmyra Museum

classification has been somewhat modified since. Cynthia Finlayson has further studied the Palmyrene female portraits. Her dissertation from 1998, *Veil, Turban and Headpiece: Funerary Portraits and Female Status at Palmyra* (University of Iowa)[39] is a welcome analysis of Palmyrene society from the portraits. The Danes are following in the footsteps of Harald Ingholt, constructing a database inventory of the Palmyra portraiture and pieces that are found around the world. The project is led by Rubina Raja from the University of Aarhus. She estimates that 2,600 Palmyrene portraits are to be found internationally in various collections. Outside Palmyra the largest portion of the Palmyra funerary sculpture is held in Denmark.[40] Laser scanning is a new way to document the sculpture by creating portraiture in 3D. It can also help to find scattered pieces to their original and right places.

An interesting group of 10 heads of funerary portraits belonging to a Roman family was found cached and inserted to the foundation of a wall connected to the 1st century AD habitation beneath the Camp of Diocletian in the western part of the city. They had been sculptured from limestone and belonged to the second style of the Palmyrene portraits.[41]

As already discussed in relation to textiles, the style of the Palmyrene sculpture is a mixture of Greco-Roman styles fused with Parthian or Persian designs, and as seen in the folds of the garments, there is even a Buddhist influence from China – no wonder the Persian styles were influential as far as Petra, as the caravan contacts to the Far East were very lively. The style in portraiture in Late Antiquity becomes less realistic compared to the Hellenistic and Imperial Roman styles; persons become stylised, even a bit simplified in their features and their posture is *en face,* namely in direct position.

The garments would have also communicated social status. Dressing largely creates the habitus of a person. Priests were obviously held in high esteem as their portraiture is so imposingly well-represented and frequent in the funerary steles and sculpture in and around Palmyra. Priests do not have a beard but a hat like a top hat resembling a Greek *polos* (Fig. 4.20, 4.23), whereas ordinary men often did not have any hat, top hat, a *modius* or turban and often have a beard in portraits (Figs. 4.19, 4.22). Rich people and dignitaries could wear a diadem. Wealthy women wore diadems or tiaras, bands, turbans, or veils (Figs. 4.17, 4.18). In Cynthia Finlayson's view the headpiece identified as a 'tiara' was actually made from textiles. Headdress types were unique to Palmyra, but influences came as far as from India.[42] Jewellery was plentiful. If wealthy, women wore a number of earrings and necklaces in a very Oriental way like Bedouin and Berber women today.

As seen already, embroidered tunics and trousers were a common outfit showing Persian or Parthian influence, and the drapery of the garments sometimes appeared to have been rendered very sharply (Ch. 2). In the funerary frescoes of Palmyra the individuals are often presented inside medallions, for example at the Tomb of the Three Brothers. The influences of exchange in the style of fresco portraiture have been seen to have reached Egypt, and some funerary customs, such as mummification, seem to have come to Palmyra from Egypt as well (see Ch. 9). Similarities in frescoes can be seen between the famous Fayum portraits in Egypt that are from the Roman period and contemporary with the art of Palmyra. In Palmyra, however, fresco portraits appear on walls and are not found as panel paintings and on the cartonnages of mummies like in

Egypt. Anyway, the mixture of Greek culture with the Orient is visible in the presentation of the people. In Palmyra portraiture on sarcophagi and funerary slabs is cut in stone in relief. Interestingly, as previously mentioned, before ISIS captured the city many heads of sculpture were already missing in Palmyra, probably due to an earlier iconoclastic rage (see Ch. 1). Some skeletal studies of deceased found in the tombs of Palmyra have been carried out. Generally Palmyrene people were short and belonged to the morphotypes typically appearing between Western Europe and East Asia. In the analyses of health conditions bone fractures and arthritis were identified, and arthritis seems to have been a common state. Surprisingly teeth were in good condition because of the fluoride in the local water.[43]

## 4.5 Private worlds of ancient Palmyrenes

There is not much published evidence of the private spaces of the ancient Palmyrenes inside the city dating to the Roman times. Beneath the Camp of Diocletian several private houses with small finds dating from the 1st century AD of the Roman period have been revealed.[44] The previously mentioned finds from the excavations by Andreas Schmidt-Colinet elucidate the Hellenistic quarters including artefacts and kinds of foods that were used by the people of Palmyra. The houses that have been in the focus of studies in Palmyra generally consist of wealthy peristyle houses and villas of the Greek and Roman type, some with frescoes and mosaics inspired by Greek mythology. Peristyle houses (Fig. 4.24) appear in various parts of the city, mainly both to the south of the Grand Colonnade as well as to the north. Some of them date from the 2nd century AD.[45] Albert Gabriel, who excavated these types of private houses in Palmyra in the 1920s, documented the plans of some houses (see Figs. 4.25–4.27).[46] They are reminiscent of the houses excavated in Pompeii in Italy with a central courtyard surrounded by columns (Fig. 4.28). In Palmyra Corinthian columns were preferred. One peristyle house next to the *Tetrapylon* served as a *Caesareum*, dedicated to the cult of the Roman emperor.[47]

A wealthy *domus* house with rich mosaics has also been unearthed behind the Temple of Bel in the eastern part of the city in the orchards. It is called the House of Cassiopeia

Fig. 4.24 *A peristyle house from Palmyra. Photographed 1920–1933.* Library of Congress, Matson Collection

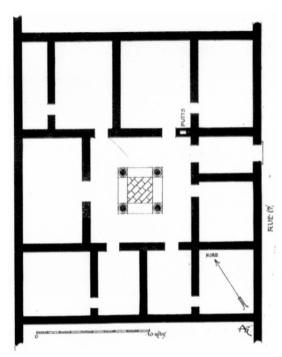

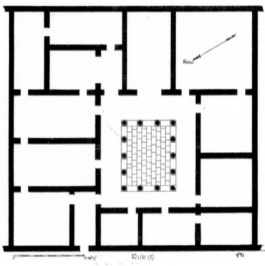

Figs. 4.25–4.27 *Types of peristyle houses that belonged to wealthy Palmyrenes.* Source: Albert Gabriel 1926

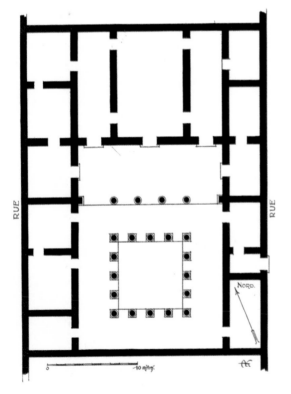

because of its mosaics inspired by the Greek story of Cassiopeia. Another house known as the House of Achilles also entails mosaics that depict the hero from the Trojan war. Akhilles was a preferred heroic figure that was also illustrated in the tombs of Palmyrenes.[48] Greek mythology penetrated into the culture and beliefs of Palmyrenes, and the beliefs of the local Semitic world were mixed with the gods and heroes of the Greco-Roman world. The mosaics of Palmyra have been compared with the famous Roman mosaics from Zeugma on the Euphrates.[49]

A fine room, probably a dining area used for banqueting and decorated with rich mosaics was unearthed in 2003 along the Grand Colonnade of Palmyra, on its northern side between the *Tetrapylon* and the so-called Funerary Temple. These symmetrical mosaics are unique. They illustrate a helmeted man combating a tiger with an arrow, and another central military looking figure 'Bellerophon'

Fig. 4.28 *A finely preserved peristyle house, the House of the Vettii, with frescoes from Roman Pompeii in Italy destroyed in AD 79 by the eruption of Vesuvius. Photographed in 1890. Library of Congress*

Fig. 4.29 *A Greco-Roman mosaic displaying god Asclepios with a Greek text as was displayed in the Palmyra Museum. Photo Kenneth Lönnqvist 2004, SYGIS, Courtesy of the Palmyra Museum*

has been symbolically identified with King Odenathus, Queen Zenobia's husband, galloping with a Pegasus-like winged horse and fighting with a javelin against a lion that apparently is symbolizing Persians. Both men in the mosaics wear Persian style of Palmyrene clothing with trousers. The central scenes are surrounded by vegetal or geometric ornamental bands in zones, sometimes with masks inserted, and surrounded by smaller figurative panels with images of various animals and plants. The colours are bright but greenish in general. Aramaic texts have been inserted into the figurative mosaics. Persian influence has been traced in the themes. The mosaics apparently date from the 3rd century AD.[50]

Fig. 4.30 *Roman pottery from Palmyra.*
Photo: Kenneth Lönnqvist 2008, SYGIS,
Courtesy of the Palmyra Museum

Fig. 4.31 *Roman pottery from Palmyra.*
Photo: Kenneth Lönnqvist 2008, SYGIS,
Courtesy of the Palmyra Museum

Fig. 4.32 *Roman glass from Palmyra.*
Photo: Kenneth Lönnqvist 2008, SYGIS,
Courtesy of the Palmyra Museum

Fig. 4.33 *Glass bracelets in the Palmyra
Museum, the types have been used from
the Roman period to the Islamic times.*
Photo: Kenneth Lönnqvist 2008, SYGIS,
Courtesy of the Palmyra Museum

In the Camp of Diocletian barracks and storehouses for soldiers appear in the Late Roman period, when Palmyra was transformed to a garrison city, and even a latrine has been identified in the ruins. In that area remains of the houses from the Arab occupation were found as well.[51] In the neighbourhood of the Agora Byzantine and Arab houses with a pottery workshop have also been revealed in excavations.[52]

Tombs of various types found in Palmyra can also reveal the private worlds of the Palmyrenes (see Ch. 9), especially *mausolea* – tombs that are like houses. As already seen, the banquet scenes in the funerary sculpture depict furniture such as *klinai*, couches, cushions, and dishes that were used. As seen, dishes of various Greco-Roman types and materials have been unearthed in excavations. Roman pottery and glass are well-presented; glass was used in jewellery as well (Figs. 4.30–4.33). Metal was also used as a material in dishes as well as in jewellery, belts, and weaponry, and we know that Palmyra was involved in the trade of metal dishes. Oil lamps that have been found in tombs tell about the light that gave visibility in the nights and darkness.[53]

## 4.6 From early Christians to Islam, Arabs, and Jews

The influence of Christianity on Palmyra must have been very early as far as the evidence of the early 3rd century AD home church in Dura Europos, a neighbouring city on the Euphrates,[54] is concerned. Queen Zenobia of Palmyra also had connections with the Christians of the East in the 3rd century AD, for example to the bishop Paul of Samosatha.[55] More visibly Christianity came to Palmyra in the 4th century AD, when the city had become a garrison city. The city had its own bishop and churches. The bishop participated in the Council of Nicea AD 325 that was one of the major turning points in the history of Christianity under Emperor Constantine the Great, who had converted to Christianity and started the programme to build churches throughout the East. A bishop from Palmyra also participated in the Council meeting of Chalcedon in AD 351.[56]

Before Islam the area was under the Yamani tribe of Kalb, the name that bears an echo to the Amorite tribal past (cfr. Bene-Yamina, namely Benjaminites – the Brothers of the Right bank of the Euphrates in the Mari texts, ARM). Later on, Arab conquests affected the city. Islam took over in the AD 630s.[57] But, as mentioned, there were Jews in antiquity as well as in the Medieval times. Some have even thought that Zenobia was a convert Jew herself, but that seems implausible, because the Palestinian Jews used to despise her and complained about her occupation of Palestine. What is known is that she had Arab roots.[58] Rabbi Benjamin Tudela reports of the inhabitants of Tadmor in the 12th century as follows (400–401):

> It (Tadmor) contains two thousand warlike Jews, who are at war with the Christians and with the Arabian subjects of Noureddin, and assist their neighbors the Mohammedans. Their chiefs are R. Isaac Hajevani, R. Nathan, and R. Usiel.

## 4.7 Village people and Bedouin society

Tadmor was just a village from Medieval times until the early 20th century. The site then concentrated on the precincts of the Temple of Bel (a phenomenon that will be further discussed in the case of the Temple of Bel and its transformations in Ch. 6) and its people must have included local *fellahin* – Arab peasants, who were engaged in agriculture and animal breeding from village-type houses (Figs. 4.34–4.37). Arab Bedouins (Figs. 4.38–4.42) spread from the Arabian Peninsula to the Palmyrides, the mountain chain and its steppes. They use tents while moving around with their flocks and, as previously explained, they can stay part of the year in village houses as transhumants. Various tribes have lived in the area

Fig. 4.34 *A Palmyrene man photographed in 1867 sleeping on a large capital in front of the Triumphal Arch.* Photo: Library of Congress, Maison Bonfils, Beirut, Lebanon

Fig. 4.35 *The oasis village of Tadmor photographed in 1920–1930.* Photo: Library of Congress, Matson Collection

Fig. 4.36 *Children from Tadmor/Palmyra in 1929.* Photo: Library of Congress, Matson Collection

Fig. 4.37 *Women and children from Tadmor/Palmyra in 1929*. Photo: Library of Congress, Matson Collection

Fig. 4.38 *Bedouins of Syria.* Photo: Library of Congress

Fig. 4.39 *Bedouins of Syria photographed in 1880*. Photo: Library of Congress, Carpenter collection

headed by chiefs or sheikhs (Fig. 4.42). The Arabic word for nomadism is *badāwa*, referring to the origin of the word for a Bedouin, meaning that Bedouins are people who went to *al-bādiya*, the desert.[59] There have been times of tension and times of peace between agriculturalists and Bedouins. The pacification of Palmyra and the surrounding desert-steppe took place in 1870;[60] however, A. Musil describes dangers in moving in the region at the beginning of the 20th century.[61]

Fig. 4.40 *A Bedouin youth from Palmyra photographed in 1920–1933*. Photo: Library of Congress, American Colony collection

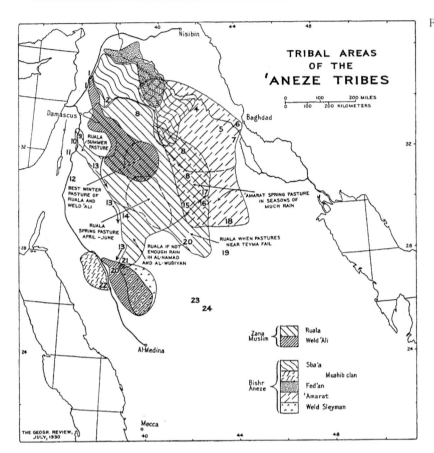

Fig. 4.41 *Tribal areas of the 'Aneze tribes.*

Gertrude Bell returned to Palmyra from the desert and wrote on April 25, 1914:

> ... beautiful gracious heights, the mountain range of Palmyra

> ...Finally we hove up against tents and camel herds of the Sba (Anazeh of Fahd Beg's people) and the herdsman's set us right. We were in fact within sight of Palmyra and I can see the bay of desert wherein it lies from my tent. For we have not reached Bukharra – I don't think we should have reached it even if we had gone straight to it. Palmyra from the desert – it must be nearly 10 miles from us – is very different Palmyra from the city you come along the Roman road from Damascus. It is very different in spirit. One looks here upon the Arab Palmyra, facing the desert, ruler of the desert and dependent upon the desert for its life and force. I am wrong to call it Palmyra; that was its bastard Roman name. Tudmor, Tudmor of the Wilderness. And the Sba know it by no other name.

The Sba'a of the Aneze confederation has indeed been the most prolific tribe in the Palmyra area (see Fig. 4.41).[62] Palmyra is especially their area for winter grazing.[63] Other tribes are the Rwala/Ruala and Weld Ali (see map). Sheep and camels are the main livelihood of

Fig. 4.42 *A Bedouin chief from Palmyra photographed in 1890-1900.* Library of Congress, Collection Views of the Holy Land

these semi-nomads, and there are territorial grazing grounds allocated to the tribes. In 1956 Al-Sba'a al-Btaynat consisted of 6,000 individuals and had a territory of 19,450 dunams, Al Sba'a al-'Abda consisted of 12,000 individuals and covered the area of 100,965 dunams. In the 1860s the Rwala had 12,000 tents in Northern Arabia.[64] The project SYGIS led by the present principal author studied the lifestyle of the transhumant Sba'a tribe in the region and the impact of desertification in their lives before the civil war.[65]

The village economy has belonged to these pastoral nomads of the transhumance type, in which part of the year is spent in highlands and part in lowlands. There is a semi-sedentary element in the lifestyle in which tented camps are used at the time of grazing and village houses with shifting agriculture are used part of the year. It is, however, apparent that inhabitants of desert areas have been more nomadic, whereas near the Euphrates and the oases the sedentary lifestyle with small livestock rearing has been adopted. Based on the observations and studies of the present author the Fi'dan tribe of the Aneze confederation uses more dromedaries in the Syrian Desert than do the Sba'a. The Fi'dan seems to be more nomadic as well.[66]

## Endnotes

1    Khazanov, A. (1994) *Nomads and the Outside World*, Madison, Wisconsin: University of Wisconsin Press, pp. 202–203.

2    See, for example, As'ad, Kh. and Yon, J.-P. (2001) *Inscriptions de Palmyre: Promenades épigraphiques dans la ville antique de Palmyre,* Guides archéologiques de l'IFAPO No. 3. Beyrouth–Damas–Amman: Institut français d'archéologique du Proche-Orient.

3    As'ad, Kh. and Yon, J.-B. (2001*) Inscriptions de Palmyre, Promenades épigraphiques dans la ville antique de Palmyre,* Guides archéologiques de l'IFAPO No.3, Beyrouth–Damas–Amman: Institut français d'archéologique du Proche-Orient, p. 96.

4    See Sakala LaBianca, Ø. (1990) *Sedentarization and Nomadization: Food System Cycles at Hesban and Vicinity in Transjordan.* Hesban 1. Berrien Springs, MI: Andrews University Press; Cribb, R. (1991) *Nomads in Archaeology.* New Studies in Archaeology. Cambridge: Cambridge University Press; Finkelstein, I. (1995) *Living on the Fringe: the Archeology and History of the Negev, Sinai and Neighbouring Regions in the Bronze and Iron Ages. Monographs in Mediterranean Archaeology.* Sheffield: Sheffield Academic Press.

5    Silver, M. (2016) *Visualising Invisible Nomads,* an invited paper presented in the World Heritage Strategy Forum at Harvard University, USA, 9–11. Sept., 2016.

6    Kupper, J.-R. (1957) *Les nomades en Mésopotamie au temps des rois de Mari.* Bibliothèque de la Faculté de Philosophie et Lettres de l'Université de Liège. Fascicule CXLII. Paris: Les Belles Lettres; see also Buccellati, G. (1966) *The Amorites of the Ur III Period.* Pubblicazioni del seminario di semistica a cura di Giovanni Garbini. Richerche I. Naples: Istituto Orientale di Napoli; see further Lönnqvist, M. (2000) *Between Nomadism and Sedentism: Amorites from the Perspective of Contextual Archaeology.* Helsinki: Juutiprint and Nilhamn, B. (2008) Nomadic Life in Central and Eastern Syria: A Perspective from the Present Life on Badiyah to the Amorite Nomadism in the Bronze Age, in *Jebel Bishri in Context: Introduction to the Archaeological Studies of Jebel Bishri and its Neighbourhood in Central Syria,* Proceedings of a Nordic Research Training Seminar in Syria, May 2004, ed. by Lönnqvist, M., BAR International Series 1817, Oxford, UK: Archaeopress, pp. 15–30.

7    Lönnqvist, M. (2010) How to Control Nomads? A Case Study Associated with Jebel  Bishri in Central Syria,

in *City Administration in the Ancient Near East, Proceedings of the 53e Rencôntre Assyriologique Internationale, Babel und Bibel 5*, ed. by L. Kogan et al., Winona Lake, Indiana: Eisenbrauns, pp. 115–139.

8    See Klengel, H. (1996) Palmyra and International Trade in the Bronze Age: the Historical Background, in *Les Annales Archéologiques Arabes Syriennes*, Vol. XLII, 1996, *Special Issue Documenting the Activities of the International Colloquium Palmyra and the Silk Road*, pp. 159–163.

9    Klengel, H. (1996) Palmyra and International Trade in the Bronze Age: the Historical Background, in *Les Annales Archéologiques Arabes Syriennes*, Vol. XLII, 1996, *Special Issue Documenting the Activities of the International Colloquium Palmyra and the Silk Road*, pp. 159–163, especially p. 160.

10    Heltzer, M. (1981) *The Suteans*, with a contribution by Arbeli, S., Istituto Universitario Orientale, Seminario di Studi Asiatici, Series Minor XIII, Naples: Istituto Universitario Orientale,  p. 25; see especially Ziegler, N. (2014) The Sutean Nomads in the Mari Period, in *Settlement Dynamics and Human-Landscape Interaction in the Dry Steppes of Syria*, Studia Chaburensia, Vol. 4, ed. by Morandi Bonacossi, D., Wiesbaden: Harrassowitz Verlag, pp. 209–226; Lönnqvist, M., Törmä, M., Lönnqvist, K. and Nuñez, M. (2011) *Jebel Bishri in Focus: Remote sensing, archaeological surveying, mapping and GIS studies of Jebel Bishri in central Syria by the Finnish project SYGIS*, BAR International Series 2230, Oxford: Archaeopress, pp. 135–212, especially pp. 202–204; the term *humusûm* p. 244, originally published by J.-M. Durand, and identified with cairns/*tumuli* by M. Durand, R. Ziegler and the present principal author M. Silver (former Lönnqvist) in 2008 after her project had surveyed cairns/*tumuli* in the region of Jebel Bishri; a former member of the Jebel Bishri project K. J. Hesse is following these original studies in her article Palmyra, Pastoral Nomads, and City-State Kings in the Old Babylonian Period: Interaction in the Semi-Arid Syrian Landscape, in *Palmyrena: City, Hinterland and Caravan Trade between Orient and Occident, Proceedings of the Conference Held in Athens, December 1–3, 2012*, ed. by Meyer, J.C., Seland, E.H. and Anfinset, N., Oxford: Archaeopress, pp. 1–9.

11    Charpin, D. (2010) The Desert Routes around Djebel Bishri and the Sutean Nomads according to the Mari Archives, in *Formation of Tribal Communities: Integrated Research in the Middle Euphrates, Syria, Al-Rafidan, Special Issue 2010*, pp. 239–245.

12    Shavit, E. (2013) The History of Desert Truffle Use, in *Desert Truffles: Phylogeny, Physiology, Distribution and Domestication,* ed. by Kangan-Zur, V., Roth-Bojenrano, N., Sitrit, Y. and Morte, A., Soil Biology, Berlin–Heidelberg: Springer pp. 214–241; Shavit, E. (2008) Truffles Roasting in the Evening Fires, in *Fungi*, Vol. 1, pp. 18–23; see the truffle habitat of the Palmyra region, especially Jebel Bishri, in Lönnqvist, M., Törmä, M., Lönnqvist, K. and Nuñez, M. (2011) *Jebel Bishri in Focus: Remote sensing, archaeological surveying, mapping and GIS studies of Jebel Bishri in central Syria by the Finnish project SYGIS*, BAR International Series 2230, Oxford: Archaeopress, pp. 62, 143 and 345; the Sumerian text called 'The Myth of Martu' or 'The Marriage of Martu' tells that the Amorites dig truffles in the foot of a mountain, see CBS 14061 in Chiera, E. (1924) *Sumerian Religious Texts*. Crozer Theological Seminary Babylonian Publications. Vol. I. Upland. PA: Crozer Theological Seminary.

13    Ziegler, N. (2004) Samsi-Addu et la Combine Sutéenne, in *Nomades et sédentaires dans le Proche-Orient ancien, Compte rendu de la XLVIe Rencôntre Assyriologique Internationale (Paris, 10-13 juillet 2000), Amurru 3*, ed. by Nicolle, C., Paris: Éditions Recherche sur les Civilisations, pp. 95–109; Charpin, D. (2010) The Desert Routes around Djebel Bishri and the Sutean Nomads according to the Mari Archives, in *Formation of Tribal Communities: Integrated Research in the Middle Euphrates, Syria, Al-Rafidan, Special Issue 2010*,  pp. 239–245, p. 243.

14    Charpin, D. (2010) The Desert Routes around Djebel Bishri and the Sutean Nomads according to the Mari Archives, in *Formation of Tribal Communities: Integrated Research in the Middle Euphrates, Syria, Al-Rafidan, Special Issue 2010*,  pp. 239–245; Lönnqvist, M., Törmä, M., Lönnqvist, K. and Nuñez, M. (2011) *Jebel Bishri in Focus: Remote sensing, archaeological surveying, mapping and GIS studies of Jebel Bishri in central Syria by the Finnish project SYGIS*, BAR International Series 2230, Oxford: Archaeopress, pp. 135–166; 202–204.

15    See Brinkman (1968) *A Political History of Post-Kassite Babylonia 1158–722 B.C.,* Analecta Orientalia 43, Commentationes scientificae de Rebus Orientis Antiquitis, Roma: Pontificium Institutum Biblicum, pp. 267–287; see also Heltzer, M. (1981) *The Suteans*, with a contribution by Arbeli, S., Istituto Universitario Orientale, Seminario di Studi Asiatici, Series Minor XIII, Naples: Istituto Universitario Orientale, pp. 94–95; see more recently Bodi, D. (2016) Is There a Connection between the Amorites and the Arameans?, in *ARAM*, Vol. 26, 1 & 2, (2015), pp. 383–409.

16    See Dhorme, P. (1924) Palmyre dans les textes assyriens, in *Revue biblique*, Vol. 33, pp. 106–108; see also Klengel, H. (1996) Palmyra and International Trade in the Bronze Age: the Historical Background, in *Les Annales Archéologiques Arabes Syriennes*, Vol. XLII, 1996, *Special Issue Documenting the Activities of the International Colloquium Palmyra and the Silk Road*, pp. 159–163, especially p. 161

17  See, for example, Izre'el, S. (1991) *Amurru Akkadian: A Linguistic Study*. With an Appendix on the History of Amurru by Singer, I. Vols. I–II. Harvard Semitic Studies 40–41. Atlanta, Georgia: Scholars Press.

18  Darmark, K. (2008) The Archaeological Potential of Assyro-Aramaean Hostility on the Euphrates Side of Jebel Bishri, in *Jebel Bishri in Context: Introduction to the Archaeological Studies of Jebel Bishri and its Neighbourhood in Central Syria*, Proceedings of a Nordic Research Training Seminar in Syria, May 2004, ed. by Lönnqvist, M., BAR International Series 1817, Oxford, UK: Archaeopress, pp. 49–58; see also Lönnqvist, M. (2010) How to Control Nomads? A Case Study Associated with Jebel Bishri in Central Syria, in *City Administration in the Ancient Near East*, Proceedings of the 53e Rencôntre Assyriologique Internationale, Babel und Bibel 5, ed. by Kogan, L. et al., Winona Lake, Indiana: Eisenbrauns, pp. 115–139.

19  Beyer, K. (1986) *The Aramaic Language: Its Distribution and Subdivision*, Göttingen: Vandenhoeck & Ruprecht, pp. 14–19.

20  Kennedy, D. (2006) Demography, the Population of Syria and the Census of Q. Aemilius Secundus, in *Levant*, Vol. 38, pp. 109–124.

21  Butcher, K. (2003) *Roman Syria and the Near East*, London: The British Museum Press, pp. 104–105.

22  Bounni, A. and Al-As'ad, Kh. (1997) *Palmyra: History, Monuments and Museum*, Damascus, p. 128.

23  Maraqten, M. (1995) The Arabic Words in the Palmyrene Inscriptions, in *ARAM*, Vol. 7, pp. 89–108.

24  Lönnqvist, M., Törmä, M., Lönnqvist, K. and Nuñez, M. (2011) *Jebel Bishri in Focus: Remote sensing, archaeological surveying, mapping and GIS studies of Jebel Bishri in central Syria by the Finnish project SYGIS*. BAR International Series 2230, Oxford: Archaeopress, p. 343.

25  Teixidor, J. (1979) *The Pantheon of Palmyra*, Leiden: Brill, p. 34.

26  Strabo, *Geography*, Book XVI, Ch. IV, transl. and ed. by Hamilton, H.C. and Falconer, W., London: George Bell & Sons, 1903.

27  Piersimoni, P. (1995) *The Palmyrene Prosopography*. Thesis submitted for the Higher Degree of Doctor of Philosophy. London: University College of London; Yon, J.-B. (2002) *Les Notables de Palmyre*, BAH, Tome 163, Beyrouth: Ifpo, p. 57–97.

28  Finlayson, C. (2002) The Women of Palmyra – Textile Workshops and the Influence of the Silk Trade in Roman Syria, in *Silk Roads, Other Roads: Textile Society of America Symposium Proceedings*, University of Nebraska – Lincoln Papers 385, pp. 70–80; see http://digitalcommons.unl.edu/cgi/viewcontent.cgi?article=1514&cont ext=tsaconf. Accessed 15th Nov. 2017.

29  Lönnqvist, M., Törmä, M., Lönnqvist, K. and Nuñez, M. (2011) *Jebel Bishri in Focus: Remote sensing, archaeological surveying, mapping and GIS studies of Jebel Bishri in central Syria by the Finnish project SYGIS*. British Archaeological Reports International Series 2230, Oxford: Archaeopress, p. 361.

30  Piersimoni, P. (1995) *The Palmyrene Prosopography*. Thesis submitted for the Higher Degree of Doctor of Philosophy. London: University College of London; see also Yon, J.-B. (2002) *Les Notables de Palmyre*, BAH, Tome 163, Beyrouth: Ifpo.

31  In Palmyra itself there are Jewish names, for example in epitaphs, see Ilan, T. and Hűnefeld, K. (2011) *Lexicon of Jewish Names in Late Antiquity*. Part IV. The Eastern Diaspora 330 BCE–650 CE. Tübingen: Mohr Siebeck; Avigad, N. (1976) *Beit Shearim*, III. Report on the Excavations during 1953–1958. New Brunswick: Rutgers University Press; see more in Teixidor, J. (2005) Palmyra in the Third Century, in *A Journey to Palmyra: Collected Essays to Remember Delbert R. Hillers*, ed. by Cussini, E., Culture and History of the Ancient Near East, Vol. 22, Leiden: Brill, pp. 181–226.

32  Yon, J.-B. (2002) *Les Notables de Palmyre, Études d'histoire sociale*, BAH, Tome 163, Beyrouth: Ifpo, p. 57–97.

33  See a collection of funerary portraits from Palmyra in Sadurska, A. and Bounni, A. (1994) *Les sculptures funéraires de Palmyre en collaboration avec Khaled al-Ass'ad et Krzysztof Makowski*. Rivista di Archeologia, diretta da Traversari, G., Roma: Giorgio Bretschneider Editore.

34  Yon, J.-B. (2002) *Les notables de Palmyre. Études d'histoire sociale*. BAH 163. Beyrouth: Ifpo.

35  Al-As'ad, Kh. and Gawlikowski, M. (1997) *The Inscriptions in the Museum of Palmyra, A Catalogue*, Palmyra and Warsaw: The Committee for Scientific Research, Warsaw, Poland, p. 21.

36  Gregoratti, L. (2015) Palmyra, City and the Territory through the Epigraphic Sources, in *Broadening Horizons 4: A Conference of young researches working in the Ancient Near East, Egypt and Central Asia*, University of Torino, October 2011, ed. by Affani, G., Baccarin, C., Cordera, L., Di Michele, A. and Gavagnin, K., BAR International Series 2698, pp. 55–59.

37  See Thompson, D.B.(1982) A Dove for Dione, in *Hesperia Supplements*, Vol. 20, Studies in Athenian Architecture, Sculpture and Topography, Presented to Homer A. Thompson, Princeton, New Jersey: American School of Classical Studies at Athens, pp. 155–162 and 215–219.

38  Ingholt, H. (1928) *Studier over Palmyrensk Skulptur* originally in Danish. Kobenhavn: C.A. Reitzels Forlag.

39   Finlayson, C. (1998) *Veil, Turban and Headpiece: Funerary Portraits and Female Status at Palmyra.* University of Iowa. Unpublished PhD. Ann Arbor, Mich.: UMI Dissertation Services, 2003. 3 v.

40   Raja, R. and Højen Sørensen, A. (2015) *Harald Ingholt and Palmyra.* Aarhus: Faellesttrykkeriet Aarhus Univeristet. We thank Professor Arja Karivieri for initially informing us about Professor R. Raja's Palmyra portrait project.

41   Michalowski, K. (1964) *Palmyre: Fouilles Polonaises 1962,* Warszawa: Editions Scientifiques de Pologne, pp. 30–32.

42   Finlayson, C. (2002–2003) Veil, turban and headpiece: funerary portraits and female status at Palmyra, in *Les Annales Archéologiques Arabes Syriennes,* Vol. XLV–XLVI, pp. 221–235.

43   See, for example, Doxiadis, E. (2000) *The Mysterious Fayum portraits: Faces from Ancient Egypt.* London: Thames & Hudson. Nakahashi, T. (2016) *On the Human Skeletal Remains Excavated from the Underground Tombs in Palmyra, in Palmyrena: City, Hinterland and Caravan Trade between Oirent and Occident, Proceedings of the Conference Held in Athens,* December 1-3, 2012, ed. by Meyer, J.C, Seland; E.H. and Anfinset, N, Oxford: Archaeopress, pp. 147-159,

44   Michalowski, K. (1966) *Palmyre: Fouilles Polonaises 1963 et 1964,* Warszawa: Editions Scientifiques de Pologne, pp. 13–14.

45   Gabriel, A. (1926) Recherches archéologiques à Palmyre, in *Syria,* Vol. 7, pp. 84–87; see also Browning, I. (1979) *Palmyra,* Park Ridge, New Jersey: Noyes Press, pp. 169–171; see also recent excavations of a residential Roman house with Hellenistic foundations in the area of Hellenistic Palmyra, see Erte, C. and Ployer, R. (2016) A Roman Residential House in the 'Hellenistic' Town of Palmyra: Archaeology, Function and Urban Aspects -Vessel Glass, in *Palmyrena: City, Hinterlad and Caravan Trade between Orient and Occident, Proceedings of the Conference Held in Athens, December 1–3, 2012,* ed. by Meyer, J.C., Seland, E.H. and Anfinset, N., Oxford: Archaeopress, pp. 93–106. Interestingly the Hellenistic phases resembled mudbrick houses of Dura Europos and showed that mudbrick as a building material was in use earlier in Palmyra after which stone was preferred in the Roman times.

46   Gabriel, A. (1926) Recherches archéologiques à Palmyre, in *Syria,* Vol. 7, pp. 84–87.

47   Genequand, D. (2008) An Early Islamic Mosque in Palmyra, in *Levant,* Vol. 40, pp. 3–15, especially pp. 7–9.

48   Stern, H. (1977) *Les mosaïques des maisons d'Achille et de Cassiopée à Palmyre.* BAH. Tome 101. Paris: P. Geuthner; see also Balty, J. (2014) *Inventaires des mosaiques dem Syrie 2: Les mosaiques de maisons de Palmyre,* BAH. Tome 206. Beyrouth: Presses d'Ifpo.

49   Balty, J. (2014) *Inventaires des mosaiques de Syrie 2: Les mosaiques de maisons de Palmyre.* BAH. Tome 206. Beyrouth: Presses d'Ifpo; see also Balty, J. (1996) Composantes classiques et orientales dans les mosaïques de Palmyre, in *Les Annales Archéologiques Arabes Syriennes,* Vol. XLII, 1996, *Special Issue Documenting the Activities of the International Colloquium Palmyra and the Silk Road,* pp. 407–416.

50   Gawlikowski, M. (2003) Palmyra, in *Current World Archaeology,* Vol. 12, pp. 27–30.

51   Michalowski, K. (1964) *Palmyre: Fouilles Polonaises 1962,* Warszawa: Editions Scientifiques de Pologne, pp. 45–46; Michalowski, K. (1966) *Palmyre: Fouilles Polonaises 1963 et 1964,* Warszawa: Editions Scientifiques de Pologne, p. 29.

52   Genequand, D. (2008) An Early Islamic Mosque in Palmyra, in *Levant,* Vol. 40, pp. 3–15, especially p. 4.

53   See metal beakers and lamps in Amy, R. and Seyrig, H. (1936) Recherches dans la nécropole de Palmyre, in *Syria,* Vol. 17, pp. 229–266; see also recent studies by Miyashita, S. (2016) The Vessels in Palmyrene Banquet Scenes: Tomb BWLH and BWRP and Tomb TYBLE, in *Palmyrena: City, Hinterlad and Caravan Trade between Orient and Occident, Proceedings of the Conference Held in Athens, December 1–3, 2012,* ed. by Meyer, J.C., Seland, E.H. and Anfinset, N., Oxford: Archaeopress, pp. 131–146.

54   Rostovtzeff, M. (ed.) (1934) *Excavations at Dura Europos,* Preliminary report of fifth season of work, October 1931–March 1932, conducted by Yale University and the French Academy of Inscriptions and Letters, pp. 238–253.

55   See, for example, Teixidor, J. (2005) Palmyra in the Third Century, in *A Journey to Palmyra: Collected Essays to Remember Delbert R. Hillers,* ed. by Cussini, E., Culture and History of the Ancient Near East Vol. 22, Leiden: Brill, pp. 181–226.

56   Genequand, D. (2008) An Early Islamic Mosque in Palmyra, in *Levant,* Vol. 40, pp. 3–15, especially p. 3 with the sources; see Constantine and the building of churches, for example, in Hunt, E.D. (1982) *Holy Land Pilgrimage in the Later Roman Empire A.D. 312–460.* Oxford: Clarendon Press.

57   Genequand, D. (2008) An Early Islamic Mosque in Palmyra, in *Levant,* Vol. 40, pp. 3–15, especially p. 3 with the sources.

58   See, for example, Teixidor, J. (2005) Palmyra in the Third Century, in *A Journey to Palmyra: Collected Essays to Remember Delbert R. Hillers,* ed. by Cussini, E., Culture and History of the Ancient Near East Vol. 22,

Leiden: Brill, pp. 181–226; see the question of Zenobia's Jewessness in Stoneman, R. (1992) *Palmyra and its Empire*, Ann Arbor: the University of Michigan Press, pp. 151–153.

59   Jabbur, J. A. (1995) *The Bedouins and the Desert*, SUNY Series in Near Eastern Studies, Albany, New York: State University of New York Press, p. 28.

60   Lewis, N.N. (1991) Taibe and El Kowm 1600–1980, in *Cahiers de L'Euphrate 5–6*, pp. 67–78.

61   Musil, A. (1928) *Palmyrena: A Topographical itinerary*. American Geographical Society. Oriental Explorations and Studies No 4, New York: American Geographical Society; Musil, A. (1927) *The Middle Euphrates: A Topographical itinerary*, American Geographical Society, Oriental Explorations and Studies No 3, New York: American Geographical Society, p. 81.

62   Wirth, E. (1971) *Syrien: Eine Geographische Landeskunde*, Wirtschafliche Ländeskunden, Band 4/5, Darmstadt: Wissenschaftliche Buchgesellschaft, Karte 11.

63   Lewis, N.N. (2009) *Nomads and settlers in Syria and Jordan, 1800–1980*, Cambridge: Cambridge University Press, p. 177.

64   Jabbur, J.A. (1995) *The Bedouins and the Desert*, SUNY Series in Near Eastern Studies, Albany: State University of New York Press, p. 527, 533.

65   See Lönnqvist, M. and Törmä, M. (2006) Observing Changes and their Effects in Nomadic Environment, Remote-sensing and Archaeology for Sustainable Development on Jebel Bishri in Syria, in *Environmental Geoinformatics and Modeling, Proceedings of ICEM '05, International Conference on Environmental Management (28–30 Oct 2005)*, Vol. IV, ed. by Anji Reddy, M., Hyderabad, India: Vedamsbooks; see also Lönnqvist, M., Törmä, M., Nuñez, M., Okkonen, J., Stout Whiting, M., Riihiaho, H. and Nissinen, M. (2010) Desertification and Ethnoarchaeology: Studying Hazards in the Nomadic Environment of Jebel Bishri in Syria, in *Proceedings of the 6th International Congress on the Archaeology of the Ancient Near East, May 5th–10th 2008, "Sapienza" – Università di Roma*, Vol. 1, Near Eastern Archaeology in the Past, Present and Future: Heritage and Identity, Ethnoarchaeological and Interdisciplinary Approach: Results and Perspectives, Visual Expression and Craft Production in the Definition of Social Relations and Status, ed. by Matthiae, P., Pinnock, F., Nigro, L., Marchetti, N. with the collaboration of Romano, L., Wiesbaden: Harrassowitz Verlag, pp. 369–390; see further Lönnqvist, M., Törmä, M., Lönnqvist, K. and Nuñez, M. (2011) *Jebel Bishri in Focus: Remote sensing, archaeological surveying, mapping and GIS studies of Jebel Bishri in central Syria by the Finnish project SYGIS*, BAR International Series 2230, Oxford: Archaeopress, pp. 371–398.

66   See Lönnqvist, M. and Törmä, M. (2006) Observing Changes and their Effects in Nomadic Environment, Remote-sensing and Archaeology for Sustainable Development on Jebel Bishri in Syria, in *Environmental Geoinformatics and Modeling, Proceedings of ICEM '05, International Conference on Environmental Management (28–30 Oct 2005)*, Vol. IV, ed. by Anji Reddy, M., Hyderabad, India: Vedamsbooks; see also Lönnqvist, M., Törmä, M., Nuñez, M., Okkonen, J., Stout Whiting, M., Riihiaho, H. and Nissinen, M. (2010) Desertification and Ethnoarchaeology: Studying Hazards in the Nomadic Environment of Jebel Bishri in Syria, in *Proceedings of the 6th International Congress on the Archaeology of the Ancient Near East, May 5th–10th 2008, "Sapienza" – Università di Roma*, Vol. 1, Near Eastern Archaeology in the Past, Present and Future: Heritage and Identity, Ethnoarchaeological and Interdisciplinary Approach: Results and Perspectives, Visual Expression and Craft Production in the Definition of Social Relations and Status, ed. by Matthiae, P., Pinnock, F., Nigro, L., Marchetti, N. with the collaboration of Romano, L., Wiesbaden: Harrassowitz Verlag, pp. 369–390; see further Lönnqvist, M., Törmä, M., Lönnqvist, K. and Nuñez, M. (2011) *Jebel Bishri in Focus: Remote sensing, archaeological surveying, mapping and GIS studies of Jebel Bishri in central Syria by the Finnish project SYGIS*, BAR International Series 2230, Oxford: Archaeopress, pp. 371–398.

# 5

# PALMYRA'S DESTINY BETWEEN ROME AND PERSIA

## 5.1 Odenathus and clashes between Rome and Persia AD 252–256

Syria had already been annexed to Rome as a province during the Hellenistic Age when Pompey conquered large areas by overcoming the Seleucids in 64/63 BC.[1] This, however, did not mean the occupation of Palmyra that was apparently already a Greek type of *polis*. General Mark Antony and his troops pillaged the site in 41 BC and found the city empty of people, as the population had left for the Euphrates. The Palmyrene archers escaped to the

Fig. 5.1 *Roads in the desert in the area of the Palmyrides between the city of Palmyra and the Euphrates. The Roman military troops crossed here and had a corridor for supplying the army.*
Photo: Minna Lönnqvist 2006, SYGIS

river to protect the territory.[2] This seems to point to the knowledge and customs of the transhumant pastoral nomadism that has been typical of the area. Movement to the Euphrates, and the Jezira in particular (an area which is attractive to nomads during drier seasons) is also supported by studies among the transhumant Sba'a tribe in the region of the Palmyrides.[3] The existence of the nomads inside and outside Palmyra has been a continuous phenomenon, and these nomads have been a source of military recruitment as well. However, tribalism has caused clashes not only between pastoral nomads and agriculturalists but between the nomads themselves and their territorial areas. Rome wished to protect its eastern frontier, and Palmyra was required to enable it to oversee its security. Caravans especially needed protection. The province of Syria included Judea, and the rule of Emperor Augustus (27 BC–AD 14) meant the time of *Pax Romana*, the Roman peace. This is when, according to the New Testament, the census took place in the time of Governor Quirinius, apparently in AD 6.[4] Palmyra is thought to have been annexed to the Roman Empire in AD 17–19,[5] but thereon it was a more or less autonomic entity under Rome. This 1st century has been seen as the establishment of *regio Palmyrena*.[6]

There was a constant rivalry between the West and the East; the areas of the Euphrates and the Tigris were areas of rivalry. Judea was a client kingdom under the province of Syria but revolted in AD 66–73.[7] Pliny the Elder (AD 23–79) describes in his *Historia Naturalis* (Natural History), how Palmyra had a destiny of its own between the empires of Rome and Parthia. During the Roman–Parthian wars the areas in Mesopotamia were conquered by Emperor Trajan in AD 115.[8] In the time of Trajan's rule, Rome gained the largest extension of its borders. Trajan employed the first Palmyrene army unit called *ala dromedariorum Palmyrenorum*,[9] referring to the use of camels. The new borders with the Parthians were set at the cities of Hatra (now in Iraq), Nisibis (modern Nusaibin in Turkey), Carrhae (Harran

in Turkey), and Edessa (modern Urfa in Turkey), and new client-kingdoms, such as Osrhoene, were created. A veteran colony was established in Carrhae (see sites in Figs. 5.3–5.5). A successful battle against the Parthians took place at Dura Europos on the Euphrates during the reign of Co-emperor Lucius Verus c. AD 165, and the area came under the domination of Rome.[10] The *Via Nova Traiana* reached from Jordan to Bostra in southern Syria, and some organisation of the *limes* must have taken place already in the area of Palmyra, as indicated by the milestones (Fig. 5.6). Emperor Hadrian (ruled AD 117–138) paid a visit to Palmyra, reinstating its role as a Roman city in the *limes* and even calling it *Hadriana Palmyra*.[11] The Palmyra Tariff has been associated with Hadrian's visit. The organisation

Fig. 5.2 *Emperor Trajan in a triumphal procession with Parthian captives.* Photo: Kenneth Lönnqvist 2008

Fig. 5.3 *The ruins from the ancient city of Nisibis, modern Nusaibin, in Turkey.*
Photo: Kenneth Silver 2016

Fig. 5.4 *The Roman city of Carrhae is situated beneath the ruins in Harran in Turkey.*
Photo: Minna Silver 2015

Fig. 5.5 *The ancient citadel of Edessa in modern Urfa in Turkey.* Photo: Kenneth Silver 2016

Fig. 5.6 *An inscribed milestone in the Palmyra Museum from the time of Trajan the elder (AD 109?).* Photo: Kenneth Lönnqvist 2004, SYGIS, Courtesy of the Palmyra Museum

Fig. 5.7 *A sculptured portrait at the Palmyra Museum has been suggested to represent Odenathus, partly because of the laurel wreath diadem.* Photo: Kenneth Lönnqvist 2006, SYGIS, Courtesy of the Palmyra Museum

of the *limes* was, however, intensified in the time of Diocletian, when the fortified *Strata Diocletiana* was created.

There are historical accounts of the role of Palmyra in the Roman wars. The city and its territories gained the status of colony in AD 203. Odenathus (ruled c. AD 260–267, see Fig. 5.7), who was first a senator by name Septimius, had become Roman consul by AD 258 and was soon the ruler of Palmyra. *The Historia Augusta*, a Roman source, has recorded Odenathus's biography in *Tyranni Triginta,* namely 'The Lives of the Thirty Tyrants or Pretenders'. In his biography Odenathus is described as a fearless warrior who was keen on hunting.

The name Odenathus is a Latinised version of the Arabic, but Odenathus's father and grandfather were apparently Arameans by origin. The historian Zosimus tells that Odenathus was a man for whose ancestors the Emperor always had great respect.[12] Odenathus later married Zenobia, who was born around AD 240. She was his second wife. From his first marriage he had a son named Hairan, known as 'Herod' in Greek, whose lifestyle followed the luxury of the Persian courts.[13]

In AD 224 the Sassanians took power from the Parthians in Persia. Despite the earlier defeats against Rome the Persians did not give up, and they soon pushed to the Euphrates.

Fig. 5.8 *Ruins of Dura Europos on the Euphrates*. Panorama: Gabriele Fangi 2010

Fig. 5.9 *The Palmyra Gate of Dura Europos*. Panorama: Gabriele Fangi 2010

There on the Euphrates stood the garrison city of Dura Europos, founded by the Seleucids in the Hellenistic Age; it had been under the Parthians until the Roman occupation that started c. AD 165. [14]

The city of Dura Europos is situated c. 200 km east of Palmyra and was a major caravan city with connections to Hatra and Palmyra. A straight desert road led from its Palmyra Gate to Palmyra. The Palmyra Gate was the main gate in the walls of Dura and faced the Syrian Desert. Three periods have been identified in the evolution of the Gate: it was originally constructed in the 2nd century BC during the Greek hegemony, when a rampart was connected to the Gate, followed by the Parthian structures and the final Roman finishing.[15] The building material at Dura was mudbrick or gypseous stone, benefiting from the clays of the alluvial zone of the Euphrates and neighbouring mountains. As already mentioned, clay was only used in the old ramparts of Palmyra dating from the Hellenistic period.

Consequently, Dura was incorporated in to the border zone on the Euphrates *limes* of the Roman Empire in the East in the 2nd and 3rd centuries AD. Excavations have unearthed from the deposits of the riverbank a thriving cosmopolitan city. Inscriptions and *papyri* tell the history of the city. The plan of the city is an example of the Hippodamian design of a grid layout following a Hellenistic ideal.[16]

More forts and fortresses were gradually built by the Romans 23–25 km from each other on the Euphrates.[17] Each had a different layout, and some had Parthian or even earlier foundations. Many of them were more or less at various times under the influence of Palmyra when it was under Rome as a *polis* and a colony. The Palmyrene archers, so famous in history, stayed in Dura Europos, and the Palmyrene cohors, *XX Cohors Palmyrenorum*, was formed there. Also a Palmyrene *strategos* appears as an administrator in Roman Dura.[18] There was a special temple for the Palmyrene gods[19] as well, and a Mithraeum[20] for the Roman soldiers who followed the cult of Mithras, the god who had absorbed features from Persia and

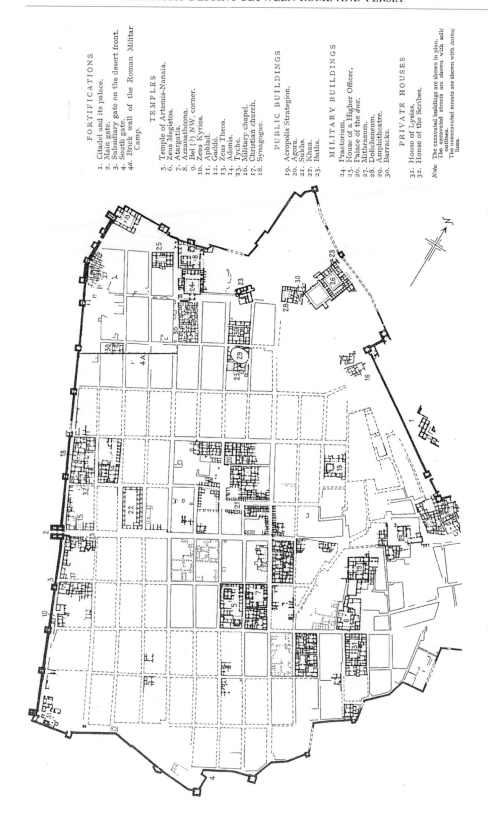

**FORTIFICATIONS**

1. Citadel and its palace.
2. Main gate.
3. Subsidiary gate on the desert front.
4. South gate.
4a. Brick wall of the Roman Military Camp.

**TEMPLES**

5. Temple of Artemis-Nanaia.
6. Zeus Megistos.
7. Atargatis.
8. Azzanathcona.
9. Bel (?) NW. corner.
10. Zeus Kyrios.
11. Aphlad.
12. Gaddé.
13. Zeus Theos.
14. Adonis.
15. Tyche.
16. Military chapel.
17. Christian church.
18. Synagogue.

**PUBLIC BUILDINGS**

19. Acropolis Strategion.
20. Agora.
21. Sukhs.
22. Khan.
23. Baths.

**MILITARY BUILDINGS**

24. Praetorium.
25. House of a Higher Officer.
26. Palace of the *dux*.
27. Mithraeum.
28. Dolicheneum.
29. Amphitheatre.
30. Barracks.

**PRIVATE HOUSES**

31. House of Lysias.
32. House of the Scribes.

*Nota:* The excavated buildings are shown in plan. The excavated streets are shown with solid outlines. The unexcavated streets are shown with dotted lines.

Fig. 5.10 *The plan of Dura Europos showing the finds revealed by the 1930s. Source: Rostavtzeff et al., 1939*

Fig. 5.11 *A reconstructed image of a Roman soldier at Dura Europos.* Courtesy: Simon James

Fig. 5.12 *The Roman fort of Qreiye on the Euphrates photographed from the air by A. Poidebard.* Source: Poidebard, 1934

Zoroastrianism.[21] In addition, as already indicated, early Christians have been proven to be in Dura by the presence of one of the earliest house-churches,[22] and a Jewish synagogue has also been discovered with colourful frescoes depicting scenes from the Jewish scriptures.[23] Dura Europos was leading a multi-ethnic as well as a multi-religious life, while being a garrison and caravan entrepôt.

At Dura Europos, Romans had to fight against Sassanians during Odenathus's reign. On the Euphrates near Dura there was the fort of Qreiye (Fig. 5.12) that was in use when Odenathus ruled. The Sassanians captured the fort c. AD 253,[24] and the neighbouring Roman garrison city of Dura underwent a long siege. Dura finally fell in AD 256. The battle was fierce, and new evidence has been discovered of the siege tactics.[25] It is apparent that, when Qreiye fell, the doom of Dura was sealed, and after the conquest by the Sassanians the site was abandoned and became desolate. Therefore, when Dura Europos was unearthed with rich finds it was vividly called 'Pompeii of the Desert'. Unfortunately, since 2011 Dura Europos has faced looting and destruction during the civil war, especially during the period it was occupied by ISIS. UNITAR and UNOSAT have traced damage and looting by satellite imagery.[26]

## 5.2 Filling the defensive gap between Palmyra and the Euphrates

In the Palmyrides, forts, fortresses, and *oppida* belonging to the Desert Limes have been identified by the project SYGIS. An early military corridor seems to have functioned in the central area of Jebel Bishri (Fig. 5.1). Villages and the base of a temporary military camp that were identified from a satellite image were traced on the ground. It provided Roman pottery, glass, and an armour scale.[27] Qseybe and Qebaqeb, Roman sites that Poidebard had earlier identified from the air,[28] were also studied on the ground. Poidebard had studied the line Soukhna–Qasr al-Her–Bir Geddid–Qebaqeb–Qseybe–Qreiye.[29] We have earlier discussed the ancient founding of the Umayyad castle of Qasr al-Hayr ash-Sharqi, and after the studies of the remains at the site we agree with A. Poidebard, M. Cary and H. H. Scullard[30] that the site was originally occupied by the Romans. We further studied Late Roman tombs in the neighbourhood; they were connected with the area of the Qasr that had several Roman architectural members in its area.[31]

In our field studies, it became clear that Qseybe was a small military post connected with Qreiye[32] that had been built in AD 208 to support Dura.[33] At Qseybe, survey finds consist of the base of a small fortlet, a well (see Ch. 3, Fig. 3.8), bases of Roman houses, pottery, and one clay *tessera* of the Palmyrene style, a Roman armour scale, and a possible Aramaic inscription on a sherd (see Figs. 5.13, 5.14).[34] As far as we know from the excavations of Qreiye, it is apparent that Romans stayed, possibly soldiers of Odenathus, at Qseybe on their way to the Euphrates before and possibly after the fall of Qreiye and the siege of Dura Europos. A larger Roman camp or fort appears on the way from Palmyra to Qseybe, namely at Qebaqeb; it is surrounded by water channels, *foggara* and *qanats*. Roman bricks, tiles, and armour scales were discovered at the Qebaqeb camp (see Figs. 5.15–5.17); however, the dating of the camp is not clear.[35]

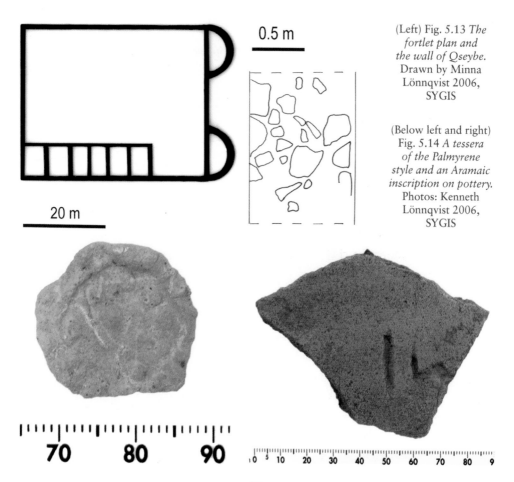

0.5 m

(Left) Fig. 5.13 *The fortlet plan and the wall of Qseybe.* Drawn by Minna Lönnqvist 2006, SYGIS

(Below left and right) Fig. 5.14 *A tessera of the Palmyrene style and an Aramaic inscription on pottery.* Photos: Kenneth Lönnqvist 2006, SYGIS

20 m

Fig. 5.15 *The valli of the Roman camp or fort of Qebaqeb between Palmyra and the Euphrates.*
Photo: Minna Lönnqvist 2006, SYGIS

Fig. 5.16 *The plan of the Qebaqeb fort or a camp.* Drawn by Kenneth Lönnqvist, SYGIS

(Above and above right) Fig. 5.17 *Roman tiles from the camp area and armour scales of Roman soldiers from Qebaqeb.* Photo: Kenneth Lönnqvist 2006, SYGIS

In light of Odenathus's career and the military, cultural, and religious connections between Palmyra and Dura Europos, one is intrigued by the question of whether Odenathus ever tried to help the Roman garrison of Dura during its siege and fall in AD 256. We posed this very question to Simon James, who has worked on the material of Dura Europos and its military strategy. He saw that, as there is no direct historical evidence, we are left with two possible scenarios to contemplate:

> 1) At the time of the siege Odenathus was still a Roman ally, and so he and his forces would have been acting under the orders of Valerian. Concentrating the largest possible field army to face Shapur (the Sassanian ruler) in pitched battle was likely a priority, and so risking piecemeal defeat of contingents was avoided.

> 2) Even if Odenathus was already ignoring Valerian, it may be that the Sassanian force concentrated to take Dura was simply too big for him to attack.

However, the Sassanians captured the Roman Emperor Valerian in Carrhae (Harran) in AD 259/Edessa in AD 260. Odenathus tried to release the Emperor and decided to fight Shapur with the Palmyrene troops. The troops encountered Shapur and drove the Persians over the Euphrates and defeated them in AD 260.[36] After the capture of Emperor Valerian, Odenathus became ruler of the East in the Roman Empire. Near the *Tetrapylon* at the Grand Colonnade of Palmyra, one column bears an Aramaic inscription referring to Odenathus, obviously posthumously. The inscription refers to Odenathus using the title of the ruler over the whole East: 'The Statue of Septimius Odenath, king and governor of the whole Orient. Septimius Zabda, the chief of a grand army, and Septimius Zabbai, the chief of Palmyra's army, have addressed this to their master' (CIS 3946). The inscription indicates that a statue stood on the bracket of the column.

Odenathus had taken part in the campaigns in Mesopotamia and fought against Persians, being victorious at Carrhae and Nisibis (modern Nusaibin in Turkey) in AD 267 (see Figs. 5.3–5.5). But when he returned he was murdered with his son Hairan (Herod) in Emesa (modern Homs) in Syria in AD 267 by a close relative. The involvement of Zenobia in the plot has been assumed, as well as other governors and Emperor Gallienus, the son of Valerian. [37]

## 5.3 Zenobia and clashes between Palmyra and Rome AD 268–272

There have been questions about the reliability of the historical accounts concerning Zenobia, Queen of Palmyra (ruled AD 267–272). The central sources from antiquity are the *Historia Augusta* and the historian Zosimus,[38] beside which there is information provided by Rufus Festus, Jordanes, Syncellus, Eutropius, Eusebius-Hieronymus, and Malalas. There are also Arabic and Jewish legends that derive from *topoi,* common themes reworked from traditional materials, such as the Book of Judith in the Bible.[39] There have been discussions among scholars on the reliability of the *Historia Augusta* as a source. There are inventions and creations of legends concerning Zenobia's biography in 'The Lives of Thirty Pretenders' in the *Historia Augusta*. Also, the 'Life of Aurelian' in the *Historia Augusta* seems to contain fabrications.[40] It needs to be remembered that the work is a Roman view. However, because of the lack of sources, accounts of Zenobia's revolt against Rome have followed closely the *Historia Augusta* and how it painted Zenobia's image

Zenobia's name was originally Iulia Aurelia Zenobia, later Septimia Zenobia; she also was called by her Aramaic name Bath-Zabbai, the daughter of Zabbai. Her father, Iulius Aurelius Zenobius, had fought against the Persians on the Roman side. As mentioned Zenobia traced her roots to the Arabs and Ptolemies of Egypt, even identifying common heritage from Cleopatra. We do not know how many children Odenathus and Zenobia had together, but they at least had one son, Vaballathus, but Herennianus and Timolaus are also mentioned in the *Historia Augusta*. Odenathus with Zenobia's stepson Hairan was murdered in Emesa in AD 267/8 after which Zenobia became regent with her son Vaballathus.[41]

Zenobia led Palmyra to revolt against Rome AD 272, 20 years after Odenathus had become ruler of Palmyra and some 16 years after Dura had fallen. During those years Zenobia had gained power and finally led Palmyra to search for its independence from Rome. She was building an empire of her own, reaching from the Nile, Palestine, and Arabia to Mesopotamia and Anatolia (Ankara) for a couple of years.[42] It is possible that Zenobia had neighbouring Dura and the Roman vulnerability in mind when she allied with the Persians. When large parts of the eastern areas of the Empire came under Zenobia's rule she minted coins decorated with her and son's portraiture. She became Augusta and her son Vaballathus Augustus. This was the propaganda of power, showing who the ruler of the East was. She was depicted on the coins wearing a helmet like a warrior queen.[43] Then Palmyra was made a capital of wealth with even more lavish architecture. Zenobia was a learned woman who spoke several languages.[44] Philosopher Longinus belonged to the court, and there were connections to Christian bishops like Paul of Samosata. Syriac authors blamed Zenobia for Paul of Samosata's fall into heresies. Several allegations of Zenobia's Jewishness appear in early Christian writings, but on the other hand there are reports how the Palestinian Jews during her rule of the East despised her. [45]

Naturally, Rome and its emperors did not approve of Zenobia's revolt against the hegemony of the empire in the East. Therefore, Emperor Aurelian decided to regain the areas under Zenobia and Palmyra. *The Historia Augusta* reports in the 'Life of Aurelian', Part 2, 22–28:

> 22 1 And so, having arranged for all that had to do with the fortifications and the general state of the city and with civil affairs as a whole, he directed his march against the Palmyrenes, or rather against Zenobia, who, in the name of her sons, was wielding the imperial power in the East. ...25 2 After this, the whole issue of the war was decided near Emesa in a mighty battle fought against Zenobia and Zaba, her ally. ....Zenobia and Zaba were put to flight, and a victory was won in full. 4 And so, having reduced the East to its former state, Aurelian entered Emesa as a conqueror, and at once made his way to the Temple of Elagabalus, to pay his vows as if by a duty common to all. 5 But there he beheld that same divine form which he had seen supporting his cause

in the battle. 6 Wherefore he not only established temples there, dedicating gifts of great value, but he also built a temple to the Sun at Rome, which he consecrated with still greater pomp, as we shall relate in the proper place.

26 1   After this he directed his march toward Palmyra, in order that, by storming it, he might put an end to his labours. But frequently on the march his army met with a hostile reception from the brigands of Syria, and after suffering many mishaps he incurred great danger during the siege, being even wounded by an arrow.

2 A letter of his is still in existence, addressed to Mucapor, in which, without the wonted reserve of an emperor he confesses the difficulty of this war: 3 "The Romans are saying that I am merely waging a war with a woman, just as if Zenobia alone and with her own forces only were fighting against me, and yet, as a matter of fact, there is as great a force of the enemy as if I had to make war against a man, while she, because of her fear and her sense of guilt, is a much baser foe. 4 It cannot be told what a store of arrows is here, what great preparations for war, what a store of spears and of stones; there is no section of the wall that is not held by two or three engines of war, and their machines can even hurl fire. 5 Why say more? She fears like a woman, and fights as one who fears punishment. I believe, however, that the gods will truly bring aid to the Roman commonwealth, for they have never failed our endeavours."

6 Finally, exhausted and worn out by reason of ill-success, he despatched a letter to Zenobia, asking her to surrender and promising to spare her life; of this letter I have inserted a copy:

7 "From Aurelian, Emperor of the Roman world and recoverer of the East, to Zenobia and all others who are bound to her by alliance in war. 8 You should have done of your own free will what I now command in my letter. For I bid you surrender, promising that your lives shall be spared, and with the condition that you, Zenobia, together with your children shall dwell wherever I, acting in accordance with the wish of the most noble senate, shall appoint a place. 9 Your jewels, your gold, your silver, your silks, your horses, your camels, you shall all hand over to the Roman treasury. As for the people of Palmyra, their rights shall be preserved."

27 On receiving this letter Zenobia responded with more pride and insolence than befitted her fortunes, I suppose with a view to inspiring fear; for a copy of her letter, too, I have inserted:

2 "From Zenobia, Queen of the East, to Aurelian Augustus. None save yourself has ever demanded by letter what you now demand. Whatever must be

accomplished in matters of war must be done by valour alone. 3 You demand my surrender as though you were not aware that Cleopatra preferred to die a Queen rather than remain alive, however high her rank. 4 We shall not lack reinforcements from Persia, which we are even now expecting. On our side are the Saracens, on our side, too, the Armenians. 5 The brigands of Syria have defeated your army, Aurelian. What more need be said? If those forces, then, which we are expecting from every side, shall arrive, you will, of a surety, lay aside that arrogance with which you now command my surrender, as though victorious on every side."

6 This letter, Nicomachus says, was dictated by Zenobia herself and translated by him into Greek from the Syrian tongue. For that earlier letter of Aurelian's was written in Greek.

28 1 On receiving this letter Aurelian felt no shame, but rather was angered, and at once he gathered together from every side his soldiers and leaders and laid siege to Palmyra; and that brave man gave his attention to everything that seemed incomplete or neglected. 2 For he cut off the reinforcements which the Persians had sent, and he tampered with the squadrons of Saracens and Armenians, bringing them over to his own side, some by forcible means and some by cunning. Finally, by a mighty effort he conquered that most powerful woman. 3 Zenobia, then, conquered, fled away on camels (which they call dromedaries), but while seeking to reach the Persians she was captured by the horseman sent after her, and thus she was brought into the power of Aurelian.

## 5.4 Zenobia – Halabiya – Zalabiya

Zosimus places Zenobia's capture on the Euphrates,[46] and traditionally the incident has been believed to have taken place near the peninsula of Halabiya, where the fortress city of Zenobia is situated (see Figs. 5.18–5.23). Tower tombs following the style of Palmyra guard the entranceways to the fortress.[47] According to Procopius,[48] the original fortress was built by Zenobia but the present one is from the time of Justinian dating from the 6th century AD.[49] Halabiya is a basalt peninsula protruding from the mountain of Jebel Bishri to the Euphrates. Throughout antiquity it was a strategically important place in the narrows of the Euphrates. On the other side of the narrows stands another fortress, known as Zalabiya, to guard the pass (see Figs. 5.19, 5.20).[50]

In the *Historia Augusta*, the 'Life of Aurelian' (Part 2, 33–35) continues with an account of Zenobia's imprisonment and the triumphal procession of Aurelian in Rome after the conquest of Palmyra:

35 4 And so Aurelian, victorious and in possession of the entire East, more proud and insolent now that he held Zenobia in chains, dealt with the Persians, Armenians, and Saracens as the needs of the occasion demanded. 5 Then were

Fig. 5.18 *The fortress of Zenobia on Halabiya.* Photo: Minna Lönnqvist 2008, SYGIS

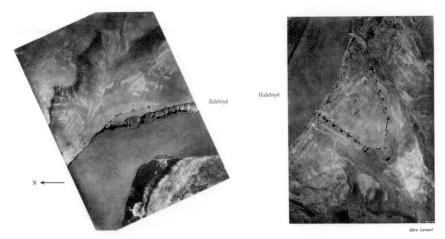

Fig. 5.20 *Zalabiya and Zenobia photographed by A. Poidebard from the air.* Source: Poidebard 1934

Fig. 5.19 *The map of Halabiya with the fortresses of Zenobia and Zalabiya by F. Sarre and E. Herzfeld 1911.*

Fig. 5.21 *F. Sarre and E. Herzfeld drew the ruins of Zenobia and published them in 1911.*

Fig. 5.22 *The fortress of Zenobia on Halabiya.* Photo: Kenneth Lönnqvist 2004, SYGIS

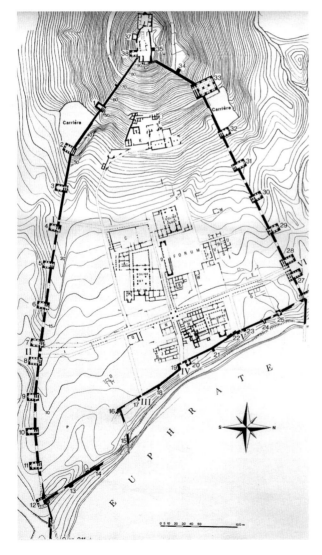

Fig. 5.23 *The plan of the fortress of Zenobia dating from the 6th century AD.* Courtesy: the Lauffray estate

Fig. 5.24 *A model of Villa Hadriana in Tivoli near Rome.* Photo: Kenneth Lönnqvist 1986

Fig. 5.25 *Pools in the gardens of Villa Hadriana in Tivoli near Rome.* Photo: Kenneth Lönnqvist 1986

brought in those garments, encrusted with jewels, which we now see in the Temple of the Sun, then, too, the Persian dragon-flags and head-dresses, and a species of purple such as no nation ever afterward offered or the Roman world beheld.

33 1 It is not without advantage to know what manner of triumph Aurelian had, for it was a most brilliant spectacle. 2 There were three royal chariots, of which the first, carefully wrought and adorned with silver and gold and jewels, had belonged to Odaenathus, the second, also wrought with similar care, had been given to Aurelian by the king of the Persians, and the third Zenobia had made for herself, hoping in it to visit the city of Rome. And this hope was not unfulfilled; for she did, indeed, enter the city in it, but vanquished and led in triumph.

34 3 And there came Zenobia, too, decked with jewels and in golden chains, the weight of which was borne by others.

There is no secure historical account of Zenobia's whereabouts in Italy but her biography in the *Historia Augusta* states that she lived with her children and was kept in her estate associated with Villa Hadriana in Tivoli (see Figs. 5.24, 5.25) near Rome.[51]

## Endnotes

1   See, for example, Sartre, M. (2005) *The Middle East under Rome*, Harvard, Cambridge, Mass.: Harvard University Press, pp. 37–40.

2   Appian, *Bella Civilia* = *The Civil Wars*, V, 9, the Loeb Classical Library 1913.

3   Lönnqvist, M., Törmä, M., Lönnqvist, K. and Nuñez, M. (2011) *Jebel Bishri in Focus, Remote sensing, archaeological surveying, mapping and GIS studies of Jebel Bishri in central Syria by the Finnish project SYGIS*, BAR International Series 2230, Oxford: Archaeopress, pp. 376–378.

4   Kennedy, D. (2006) Demograpy, the Population of Syria and the Sensus of Q. Aemilius Secundus, in *Levant*, Vol. 38, pp. 109–124.

5   Stoneman, R. (1995) *Palmyra and Its Empire: Zenobia's Revolt against Rome*, Ann Arbor: The University of Michigan Press, p. 27, some think that the annexation could already have taken place under Augustus as early as AD 14.

6   Edwell, P.M. (2008) *Between Rome and Persia: the middle Euphrates, Mesopotamia and Palmyra under Roman Control*, London and New York: Routledge, p. 41.

7   See Josephus, *Bellum Judaicum* = *The Jewish War*. The Loeb classical library 1927–. Transl. by Thackeray, H. St. J. Cambridge; Mass.: Harvard University Press; see also, for example, Smallwood, E.M. (1976) *The Jews under Roman Rule: from Pompey to Diocletian, a Study of Political Relations*, Leiden: Brill; also Rhoads, D.M. (1976) *Israel in Revolution 6-74 CE*. Philadelphia, PA: Fortress.

8   Cary, M. and Scullard, H. H. (1984) *A History of Rome*, London: MacMillan, pp. 438–439.

9   Stoneman, R. (1995) *Palmyra and Its Empire: Zenobia's Revolt against Rome*, Ann Arbor: The University of Michigan Press, p. 27.

10  Cary, M. and Scullard, H. H. (1984) *A History of Rome*, London: MacMillan, pp. 438–439.

11  Inscriptions found in Palmyra commemorate Hadrian's visit. One bilingual inscription in Greek and Aramaic was situated in the Temple of Baalshamin in Palmyra and dated to AD 131, another at the Agora of Palmyra from the same year; see, for example, As'ad, Kh. and Yon, J.-B. (2001) *Inscriptions de Palmyre, Promenades épigraphiques dans la ville antique de Palmyre*, Beyrouth–Damas–Amman: IFAPO, pp. 46–48, 60.

12  Zosimus, *Nova Historia* = *New History*, Book I, 21, London: Green & Chaplin 1814: '...Odonathus of Palmyra, a person whose ancestors had always been highly respected by the emperors... several cities belonging to the Persians, he retook Nisibis also, which Sapor had formerly taken.'

13  See *Historia Augusta, The Lives of the Thirty Pretenders*, Ch. 15 and 16: Odaenathus, Herodes (Hairan), The Loeb Classical Library 1932; inscriptions in Palmyra include an earlier Odenathus, as well as those referring to Odenathus with titles Consul (A 1247/6532) and the King of Kings, governor of the whole East (CIS 3946) and his son Hairan's (Herod) victory (IGRR III, 1032); see more in Stoneman, R. (1995) *Palmyra and Its Empire: Zenobia's Revolt against Rome*. Ann Arbor: The University of Michigan Press.

14  Edwell, P. (2008) *Between Rome and Persia: the middle Euphrates, Mesopotamia and Palmyra under Roman Control*, London and New York: Routledge, for example, p. 3.

15  Gelin, M., Leriche, P. and Abdul Massih, J. (1997) La Porte de Palmyre à Doura-Europos, in *Doura-Europos, Études IV, 1991–1993*, Beyrouth: IFAPO, pp. 20–46.

16  See, for example, Børlit, C. (2008) Dura Europos: the Final Siege and Abandonment, in *Jebel Bishri in Context, Introduction to the Archaeological Studies and the Neighbourhood of Jebel Bishri in Central Syria, Proceedings of a Nordic Research Training Seminar in Syria, May 2004*, ed. Lönnqvist, M., BAR International Series 1817, Oxford: Archaeopress, pp. 99–110.

17  See Poidebard, A. (1934) *La Trace de Rome dans le Désert de Syrie: Le Limes de Trajan a la conquête arabe. Recherches aériennes (1925–1932)*. BAH. Tome XVIII. Paris: Paul Geuthner. Texte.

18  Edwell, P.M. (2008) *Between Rome and Persia: the middle Euphrates, Mesopotamia and Palmyra under Roman Control*, London and New York: Routledge, p. 116.

19  See Dirven, L. (1999) *The Palmyrenes of Dura Europos: A Study of Religious Interaction in Roman Syria*, Leiden. Boston, Köln: Brill, pp.41–42.

20  Cumont, F. (1975) The Dura Mithraeum, in *Mithraic Studies*, Vol. I, Proceedings of the First International Congress on Mithraic Studies, Manchester, 1975, ed. by Hinnells, J.R., Manchester: Manchester University Press, pp. 151–214.

21  See, for example, Cumont, F. (1903) *The Mysteries of Mithras*. Chicago: Open Court Publishing.

22  Rostovtzeff, M. (ed.) (1934) *Excavations at Dura Europos*, Preliminary report of fifth season of work, October 1931–March 1932, conducted by Yale University and the French Academy of Inscriptions and Letters, New Haven: Yale University Press, pp. 238–253.

23  *Preliminary report of the synagogue at Dura* (1936) New Haven: Yale University Press; du Mesnil du Buisson, Comte (1936) Les deux synagogues successives à Dura Europos, in *Revue biblique*, Vol. XLV, pp. 72–90.

24  Gschwind, M. and Hasan, H. (2008) Das römische Kastell Qreiye-'Ayyāš, Provinz Deir Ez-Zor, Syrien, Ergebnisse des syrisch-deutschen Kooperationsprojektes, *Zeitschrift für Orient-Archäologie*, Vol. 1, pp. 316–334, especially p. 325.

25  See Børlit, C.S. (2008) Dura Europos: the Final Siege and Abandonment, in *Jebel Bishri in Context, Introduction to the Archaeological Studies and the Neighbourhood of Jebel Bishri in Central Syria, Proceedings of a Nordic Research Training Seminar in Syria, May 2004*, ed. by Lönnqvist, M., BAR International Series 1817, Oxford: Archaeopress, pp. 99–110; see also James, S. (2011) Stratagems, Combat, and "Chemical Warfare" in the Siege Mines of Dura-Europos, in *American Journal of Archaeology*, Vol. 115, pp. 69–101; Baird, J.A. is following the same methodological approach on the siege and abandonment of Dura originally used by Børlit, C. (2004) and published (2008) under the supervision of the present principal author M. Silver (former Lönnqvst), see: Baird, J.A. (2012) *Dura Deserta*: the Death and Afterlife of Dura-Europos, in *Vrbes Extinctae: Archaeologies of Abandoned Classical Towns*, ed. by Christie, N. and Augenti, A., Burlington: Ashgate Publishing, pp. 307–330.

26  See http://unosat.web.cern.ch/unosat/unitar/downloads/chs/Dura_Europos.pdf. Accessed 12th Oct., 2016.

27  Lönnqvist, M., Törmä, M., Lönnqvist, K. and Nuñez, M. (2011) *Jebel Bishri in Focus, Remote sensing, archaeological surveying, mapping and GIS studies of Jebel Bishri in central Syria by the Finnish project SYGIS*, BAR International Series 2230, Oxford: Archaeopress, pp. 269–284.

28  Poidebard, A. (1934) *La Trace de Rome dans le Désert de Syrie: Le Limes de Trajan a la conquête arabe. Recherches aériennes (1925–1932)*, Texte, BAH, Tome XVIII. Paris: Paul Geuthner, Texte, pp. 91–92.

29  Poidebard, A. (1934) *La Trace de Rome dans le Désert de Syrie: Le Limes de Trajan a la conquête arabe. Recherches aériennes (1925–1932)*. Texte. Atlas. BAH. Tome XVIII. Paris: Paul Geuthner, Texte, pp. 91–92.

30  Cary, M. and Scullard, H. H. (1984) *A History of Rome*, London: MacMillan, pp. 440.

31  Lönnqvist, M., Törmä, M., Lönnqvist, K. and Nuñez, M. (2011) *Jebel Bishri in Focus, Remote sensing, archaeological surveying, mapping and GIS studies of Jebel Bishri in central Syria by the Finnish project SYGIS*, BAR International Series 2230, Oxford: Archaeopress, pp. 254–261.

32  Lönnqvist, M., Törmä, M., Lönnqvist, K. and Nuñez, M. (2011) *Jebel Bishri in Focus, Remote sensing, archaeological surveying, mapping and GIS studies of Jebel Bishri in central Syria by the Finnish project SYGIS*, BAR International Series 2230, Oxford: Archaeopress, pp. 284–293.

33  Gschwind, M. and Hasan, H. (2008) Das römische Kastell Qreiye-'Ayyāš, Provinz Deir Ez-Zor, Syrien, Ergebnisse des syrisch-deutschen Kooperationsprojektes, in *Zeitschrift für Orient-Archäologie*, Vol. 1, pp. 316–334, especially p. 323.

34   Lönnqvist, M., Törmä, M., Lönnqvist,K. and Nuñez, M. (2011) *Jebel Bishri in Focus, Remote sensing, archaeo-logical surveying, mapping and GIS studies of Jebel Bishri in central Syria by the Finnish project SYGIS*, BAR International Series 2230, Oxford: Archaeopress, pp. 284–291

35   Lönnqvist, M., Törmä, M., Lönnqvist, K. and Nuñez, M. (2011) *Jebel Bishri in Focus, Remote sensing, archae-ological surveying, mapping and GIS studies of Jebel Bishri in central Syria by the Finnish project SYGIS*, BAR International Series 2230, Oxford: Archaeopress, pp. 284–291

36   See, for example, Stoneman, R. (1995) *Palmyra and Its Empire: Zenobia's Revolt against Rome*. Ann Arbor: The University of Michigan Press, pp. 76–79.

37   See, for example, Stoneman, R. (1995) *Palmyra and Its Empire: Zenobia's Revolt against Rome*. Ann Arbor: The University of Michigan Press, pp. 76–79, especially pp. 106–109.

38   Zosimus, *Nova Historia = New History*, Book I, 21, London: Green & Chaplin 1814: '.. several cities belonging to the Persians, he retook Nisibis also, which Sapor had formerly taken..'

39   See, for example, Weststeijn, J. (2013) Zenobia of Palmyra and the Book of Judith: Common Motifs in Greek, Jewish, and Arabic Historiography, in *Journal for the Study of the Pseudepigrapha*, Vol. 22, pp. 295–320.

40   See, for example, Burgersdijk, D. (2004–2005) Zenobia's Biography in the Historia Augusta, in *TALANTA, Proceedings of the Dutch Archaeological and Historical Society*, XXXVI–XXXVII, pp. 139–152.

41   See discussions of the ancient sources, for example, *Paulys Realencyclopädie der Classischen Altertumswissenschaft*(1972) Zweite Reihe, Neuzehnter Halbband (XIX), s.v. Zenobia, col. 1–7; also *Der Neue Pauly, Enzyklopädie der Antike* (2003), Band 12/2, Stuttgart, Weimar: J.B. Metzler, s.v. Zenobia, col. 730–734.

42   See, for example, Stoneman, R. (1995) *Palmyra and Its Empire: Zenobia's Revolt against Rome*. Ann Arbor: The University of Michigan Press, pp. 111–124.

43   See, for example, Schwentzel, C.-G. (2010) La Propagande de Vaballath et Zénobie d'après des Témoignages des Monnaies et Tessères, in *Rivista Italiana di Numismatica*, Vol. CXI, pp. 157–172; several series of coins were issued by Zenobia and Vabalathus in Antioch, Syria and in Alexandria, Egypt, including the title of Augustus: see Bland, R. (2011) The Coinage of Vabalathus and Zenobia from Antioch and Alexandria, in *Numismatic Chronicle* 171, pp. 133–186.

44   See, for example, Stoneman, R. (1995) *Palmyra and Its Empire: Zenobia's Revolt against Rome*. Ann Arbor: The University of Michigan Press, pp. 111–124.

45   Teixidor, J. (2005) Palmyra in the Third Century, in *A Journey to Palmyra: Collected Essays to Remember Delbert R. Hillers*, ed. by Cussini, E., Culture and History of the Ancient Near East, Vol. 22, Leiden: Brill, pp. 181–226.

46   Zosimus, *Nova Historia = New History*, Book I, 27, 28, London: Green & Chaplin 1814

47   See, for example, Blétry, S. (2012) Les nécropoles de Halabiya-Zénobia, Premier resultats (2009 et 2010), in *Syria*, Vol. 89, pp. 305–330.

48   Procopius, *Buildings*, Book 2, 8, the Loeb Classical Library 1940.

49   See Lauffray, J. (1983) *Halabiyya-Zénobia: place forte du Limes oriental et la Haute-Mésopotamie au VI siècle*. Vol. I. Les duchés frontaliers de Mésopotamie et les fortifications de Zénobia. Institut français d'archéologie du Proche-Orient. BAH. Tome 114. Paris: Librarie orientaliste Paul Geuthner; see also more recent publication by Blétry, S. dir. (2015) *Zénobia-Halabiya, habitat urbaine et nécropoles: Cinq années de recherches de la mission syro-française (2006–2010)*. Montpellier: Presses Universitaires de la Méditerrannée.

50   See ancient sources on the fortress of Zenobia and Zalabiya, for example, in Musil, A. (1927) *The Middle Euphrates, A Topographical Itinerary*, American Geographical Society Oriental Explorations and Studies No. 3, New York: The American Geographical Society, pp. 331–333.

51   *Historia Augusta, The Lives of the Thirty Pretenders*, Zenobia, Ch. XXX, 27, The Loeb Classical Library 1932: 'Her life was granted her by Aurelian, and they say that thereafter she lived with her children in the manner of a Roman matron on an estate that had been presented to her at Tibur, which even to this day is still called Zenobia, not far from the palace of Hadrian'.

# 6

# THE TEMPLE OF BEL AS THE CORE OF PALMYRA

## 6.1 The religion of Palmyra in Greco-Roman times

The Palmyrenes were strikingly religious people, a fact that is reflected in a number of cultic spaces in the city – temples, sanctuaries, shrines, and funerary structures – as well as in inscriptions, altars, statues, or reliefs depicting deities, priests, and worshippers (Figs. 6.1–6.6). Several altars were dedicated to an unknown god. The Palmyrenes were polytheists, worshipping numerous gods, but their gods had a special hierarchy.

There were early gods which had a tribal origin, such as the Arab goddess Allat and Baalshamin, the Lord of Heavens, with West Semitic Canaanite/Phoenician origins. The ancestry of these gods which had temples in Palmyra may be traced to the Hellenistic city.[1] The Palmyrene god 'Arsû also had associations with the Canaanite god Reseph,[2] and also with Azizos.[3]

The religion of the Palmyrenes was basically Semitic and Syro-Mesopotamian in nature. It had similarities with Hatra, but also many differences.[4] In Palmyra the main triad consisted of Bel, Yarhibol, and Aglibol. In AD 32, Bel as a chief deity was given the major temple, which followed an earlier structure. It is commonly held that the god was Marduk Bel, the god who belonged to the Babylonian and thus Mesopotamian pantheon. The festival of Nisan – the New Year festival in Babylon – was also celebrated in Palmyra.[5] This was partly reinstated by the earlier sanctuary found at the site of the Temple of Bel and dating from the 1st century BC, with its family gods headed by Bel.[6] Yarhibol corresponded with the Sun and Aglibol the Moon in astral beliefs. The female goddess Ishtar, a common Mesopotamian goddess, was known in Syria-Palestine as Astarte, and she was added into the deities of Palmyra as well. She also had astral features.[7]

The trapezoidal temple beside the Grand Colonnade of Palmyra was dedicated to Nabu or Nebo (see Ch. 8), son of Marduk Bel; this site has been also used as evidence of the Mesopotamian origins of Bel in Palmyra.[8] There have, however, been discussions about the evolution of the god and whether Bel was originally same as the Syro-Palestinian Baal, the storm and weather god, more specifically appearing in the dedicated Temple of Baalshamin,

Figs. 6.1–6.3
*Reliefs from
the Temple of
Bel illustrating
worshippers
and rituals that
took place at
the site in the
Roman period.*
Photos 6.1 and
6.2: Gabriele
Fangi 2010.
Photo 6.3:
Syria Journal,
1934

another important sanctuary in Palmyra. The Semitic names Bel, Baal, and Bol, translated as 'Lord,' have occurred in personal private names or family names in Palmyra and Syria-Palestine in general for a few thousand years. The tomb of Elahbel is one of the examples bearing Bel in the family name (see Ch. 9). Jezebel, the princess of Tyre from Phoenicia (the present coast of Lebanon) mentioned in the Bible,[9] has inspired authors and song writers. In Palmyra there also appears a god called by the Aramaic name 'Melchibel'. The name means 'messenger of the Lord' – this word Melech still echoes in the Syriac dialect of Aramaic, meaning an 'angel,' the messenger of God.

The weather and celestial gods had a special place in the religion of Palmyra and appear to also have been particularly associated with the neighbouring cities of Palmyra: Heliopolis-Baalbek and Emesa (modern Homs) – the inhabitatants of which worshipped astral gods. Just as Bel was the chief god in Palmyra, Baalbek had the Temple of Jupiter dedicated to the Heliopolitan Baal,[10] and in Emesa the cult was dedicated to Sol Invictus, the Invincible Sun, a solar deity. In Emesa there was a famous stone that was also worshipped, and the high priest of Emesa was held in high esteem in the Roman world, particularly during the reign of Emperor Elagabalus (Heliogabalus).[11]

Reliefs presenting the triad of Bel, Aglibol, and Yarhibol have been found in Palmyra. One was situated in the Temple of Bel (Fig. 6.4), another in the Temple of Baalshamin. One is now in the Louvre in Paris. In these carved stones, celestial gods have been depicted with their own special astral symbols and are wearing the Roman military style cuirass combined with Parthian clothing; such a mixture was also expressed in the cult of Mithras among the Roman troops. The solar crown and the crescent of the moon refer to the main celestial deities. In Dura Europos, where the temple of the Palmyrene gods was discovered, there is a fresco illustrating a Palmyrene god wearing Parthian clothing. The goddess Allat, who was

Fig. 6.4
The triad of
Palmyra carved
in relief at the
Temple of Bel.
Photo: Silvana
Fangi 2010

Fig. 6.5 *Tessserae, tickets, for religious feasts. After Bounni and Al'Asad 1997.*

worshipped in Palmyra, seems to have been an Arab god transferred from Greece and its pantheon. She has been identified with Athene, the goddess of science and arts and the city goddess of Athens. In her cult there was the tendency to monotheism, worshipping one sole god (see Ch. 10).

The Palmyrenes spread their sanctuaries around the Roman world wherever their merchants happened to be stationed. Beside Dura Europos where the Temple for the Palmyrene gods was discovered, other temples of the Palmyrenes are also known in Babylon dating from the 1st century AD and Vologesias dating from the 2nd century AD. The temple in Dura dates from 32 BC; it went through some changes and was used for about 200 years by the Palmyrenes, who visited the site during their trade missions.[12] In Rome there were several temples dedicated to the Palmyrene gods as well.[13] This was the time of the growth of Oriental religions in the Roman world; the cult of Mithras from Persia had become favoured by Roman soldiers and the sanctuaries of the god specifically appeared in garrison cities.

Fig. 6.6 *The funerary couch of the priestly family Be'elai dated to AD 180.*
Photo: Kenneth Lönnqvist 200, SYGIS, Courtesy of the Palmyra Museum

Religious banquets were part of ritual feasting in Palmyra. This was a very old Oriental and local West Semitic way to worship and to remember the deceased. In the West Semitic world the association for ritual feasting was known as *marzeah*.[14] Participants gained admission to the religious banquets in Palmyra with small tickets called *tesserae* made of metal, glass, or most commonly terracotta slabs of various shapes. These entrance tickets are reminiscent of those the Romans used to enter various theatre and circus performances or festivals (even today Italians still call bus tickets *tesseras*). In Palmyra, the *tesserae* were decorated in relief and the names of the owners were inscribed on them. Apart from deities and buildings like Bel's temple the *tesserae* depict other decorative motifs including the sun, moon, stars, rosettes, or just points,[15] (Fig. 6.5) as seen in the finds of the Qseybe fortlet in the Palmyrides (Fig. 5.14). With a ticket one could enter a ritual assembly or club, *marzeah*, (Greek *thiasos*). In Palmyra during the festivals, banquets seem to have been part of the feasting. Images have been found in the Palmyrene graves of the deceased still feasting, drinking from cups, *skyphoi*, and eating in their best costumes, surrounded by relatives (see Figs. 4.1, 6.6). The funerary structures and sculptures as well as mummification show clear belief in the afterlife (see Ch. 9).

Fig. 6.7 *A panoramic view of the ritual courtyard, temenos, of the Temple of Bel. The cultic room, the cella or naos, still stands on the left.* Photo: Gabriele Fangi 2010

(Above) Fig. 6.8 *The suggested reconstruction of the* Propylaea, *the gatehouse to the courtyard of the Temple of Bel in Wood 1753.*

(Right) Fig. 6.9 *The* Propylaea *(the tall building on the right as left after fortification of the structure in Islamic times) and the* temenos *of the Temple of Bel photographed from the Triumphal Arch and the Grand Colonnade in 1900.* Library of Congress, the American Colony collection

Fig. 6.10 *The* Propylaea *building (as it appears after fortification of the site in Islamic times) photographed from inside the courtyard of the Temple of Bel in 1936.* Library of Congress, Matson Collection

## 6.2 The creation of the Temple of Bel and documentation of its structures

The Roman architect and engineer Vitruvius wrote in the 1st century BC:

> The design of a temple depends on symmetry, the principles of which must be most carefully observed by the architect. They are due to proportion, in Greek analogia ... in the members of the temple there ought to be the greatest harmony in the symmetrical relations of the different parts to the general magnitude of the whole.[16]

The Temple of Bel with its courtyard has formed a focal point in the landscape of Palmyra. As previously mentioned, the site of the temple is located on the *tell* of the original Bronze Age city. The *Propylaea* or *Propylaia* (Figs. 6.8–6.12) was a pillared gatehouse and led to the temple courtyard in Roman times, similar to the earlier sacred area of the Acropolis in Athens. The grand *Propylaea* was approached by a wide staircase and in front stood high Corinthian columns (Fig. 6. 8, 6.10, 6.12). From that gatehouse, one entered a large rectangular courtyard, namely the *temenos* or *haram*, that had a central cult room, the *cella* or *naos*, identified with the temple itself (Fig. 6.7). The *Propylaea* of the temple area in Palmyra was destroyed in AD 1132 and transformed to a citadel for a fortification system (see Figs. 6.9–6.11) during the Seljuk rule.[17]

The courtyard was roughly a square covering 205 metres × 210 metres, and the *cella* in its inner plan covered 12.56 metres × 37.70 metres, and with the surrounding columns the area covers 27.65 metres × 52.78 metres.[18] The long walls of the *cella* are aligned south to the north, and the *cella* was entered from the side of the western long wall. The Temple of Bel apparently had at least a Hellenistic predecessor at the site, the ruins of which have been traced in the area of the temple of the Roman period. The dating of the latter temple

Fig. 6.11 *The* Propylaea *building (as it appears after fortification of the site in Islamic times) photographed from inside the courtyard of the Temple of Bel in 2010.* Photo: Silvana Fangi

Fig. 6.12 *The virtual reconstruction of the* Propylaea – *suggested view of the entrance, as it may have appeared in the Roman times.* Construction by Ahmet Denker 2016

Fig. 6.13 *The side walls of the Temple of Bel seen from outside, photographed in 1867.* Photo: Library of Congress, Maison Bonfils, Beirut, Lebanon

Fig. 6.14 *The side walls of the Temple of Bel seen from outside.* Photo: Gullög Nordquist 2004, SYGIS

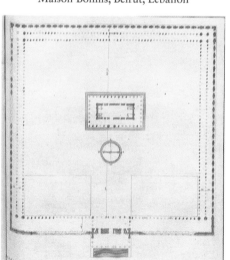

Fig. 6.15 *One of the earliest plans of the Temple of Bel and its courtyard and propylaea by Robert Wood 1753.*

to AD 32 is based on a dedicatory inscription associated with a statue that was found attached to the temple. According to the inscription, the powerful tribe of Komare had been central in the erection of the temple. The large spacious and rectangular *temenos* of the Temple of Bel was finished in the 2nd century AD.[19]

In the temple courtyard the colonnades ran on each side of the walls, the western one at the entrance rising higher than the others and changing from a double to a single row of columns, being also wider. The walls surrounding the courtyard were screened by inserted Corinthian pilasters between which there are Greek style windows with triangular pediments (see Figs. 6.13, 6.14) above the frames, similar to the Tariff Court of Palmyra and the Temple of Baalshamin (see more in Chs 7 and 8).

The plans of the temple and its courtyard were first published in Robert Wood's *The Ruins of Palmyra* in 1753 (Fig. 6.15). Betty Ratcliffe constructed a delightful miniature model of the ruins of the temple (see page 5), probably inspired by Wood's documentation (see Ch. 1). After Robert Wood and his expedition, further studies and recording were carried out by Theodor Wiegand with his team and published in 1932.[20] Furthermore, Henri Seyrig, Robert Amy, and Ernest Will[21] published in 1975 detailed documentations of the temple with its courtyard (Fig. 6.16). Since the 1930s aerial photographs have clearly displayed the *cella* surrounded by the courtyard (Figs. 6.17, 6.18). It is, however, clear that Wood's original work served as the basis of the later architectural publications. As well as the *cella* the courtyard had an altar (A) for sacrifices, a basin (B) for ablution for purification rituals, and the banquet hall (C) for ritual banquets (See Fig. 6.16). In the banquet hall several *tesserae* have been found to indicate to their use as entrance tickets to the religious banquets.[22]

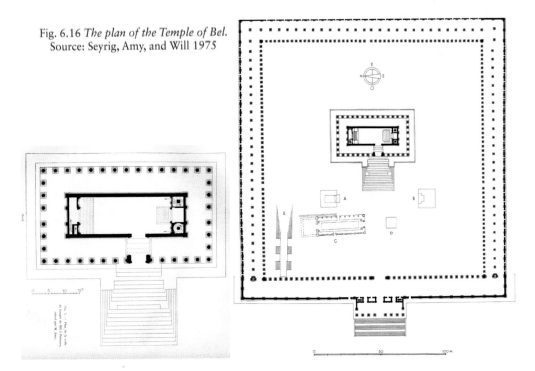

Fig. 6.16 *The plan of the Temple of Bel.*
Source: Seyrig, Amy, and Will 1975

Fig. 6.17 *An aerial photograph of the Temple of Bel taken in 1932.* Source: Poidebard, 1934

Fig. 6.18 *An aerial photograph taken of the Temple of Bel in 1962.*

Fig. 6.19 *A panorama towards the cella of the Temple of Bel taken in the vast courtyard.*
Photo: Gabriele Fangi 2010

The temple with its courtyard (Fig. 6.19) was built of limestone that Gertrude Bell viewed while walking in the *temenos* on 21 May 1900 and described in a letter: 'The stone used here is a beautiful white limestone that looks like marble and weathers a golden yellow, like the Acropolis.' In the Temple of Bel, the outer walls of the rectangular *naos* or *cella*, the

Fig. 6.20 *The portal of the Temple of Bel seen behind mud houses photographed in 1867.* Photo: Library of Congress, Maison Bonfils, Beirut, Lebanon

Fig. 6.21 *The cella of the Temple of Bel in the 1930s–1940s.* Photo: Library of Congress, Matson Collection

Fig. 6.22 *The* cella *of the Temple of Bel photographed in the 1930s–1940s.* Photo: Library of Congress, Matson Collection

Fig. 6.23 *The* cella *of the Temple of Bel.* Photo: Gabriele Fangi 2010

central cultic room, were screened with Ionian pilasters, and the whole structure was surrounded by the peristyle of Corinthian columns.

However, the temple had an Oriental touch that follows long-lasting local traditions – here old Semitic and Greek traditions were fused. As far as the rectangular layout is concerned, the Neolithic Near East had already produced the rectangular temple structure,[23] and the Greeks or Anatolians invented the

Fig. 6.24 *The cella of the Temple of Bel.* Photo: Gabriele Fangi 2010

*megaron*.[24] In Syria a rectangular temple *in antis* became a popular form for religious spaces in the Early Bronze Age.[25] This form also appears in Palestine and was preferred there in the Middle Bronze Age. An Early Bronze Age rectangular temple is known in Ebla, appearing in a tri-partite form in the Great Temple D dating to ca. 2000 BC.[26] Tell Tayinat in the Plain of Antioch (now in Turkey) shows that in the Iron Age Arameans were also interested in the rectangular design of tri-partite form comparable to the Temple of Solomon. The latter Jewish temple was built by the Phoenicians, according to the Bible[27] and has been dated to the Iron Age, 10th century BC.[28] The later temple built by Herod followed the same layout. Here in the

Temple of Bel, unlike in many Semitic counterparts, the rectangular room is a *Breitraum* or *Breithaus*, a wide room or house that is entered from the long side. In contrast, the *Langraum* or *Langhaus*, the long room type, entered from the short side, was common in Syria in the Bronze Age.[29] But what is a typical Semitic organisation of the courtyard with a *cella* was the circumambulatory plan, with concentric rectangular structures successively inside each other.[30]

In the *cella* of the Temple of Bel the pilasters screening the outer walls are of the Greek Ionian order, and the columns of the surrounding *peristyle* of the *cella* were of the Corinthian order (Figs. 6.21–6.24). The monumental western entrance portal to the temple was originally part of the *peristyle*,[31] the columns surrounding the temple, but stood apart, and was the only structure left standing when the jihadists destroyed the *cella*. The portal is suggestive of the gates to the palace of Persepolis in Persia, and some have seen features of the gateways to Ptolemaic temples in Egypt.[32] From surviving inscriptions we know that the building of the Temple of Bel involved Greek workers, and some think that the architect was also Greek, but he has also been suggested to originate from Syrian Antioch.[33] The Greek parallels have been traced to Hellenistic architect Hermogenes of Priene and the Temple of Artemis in Magnesia on the Meander. Hermogenes had influenced cultic structures not only in Anatolia but also in Syria under the Seleucids.[34] In any case, as stated above, with its *Breitraum* layout of the *cella* inside a large courtyard the temple has a clear Semitic touch. One may also see architectural features from earlier tent-shrines, such as a tabernacle.

The *cella* had several cultic areas, and at each end there was a *thalamos* or *adyton*, a kind of holy of the holies, with niches for statues of deities. The *thalamoi* were different from each other. The northern *thalamos* was taller and had plain Dorian pilasters while its southern

Fig. 6.25 *Assyrian merlons visible on the roof of the* peristyle *of the* cella.
Photo: Minna Lönnqvist 2000, SYGIS

Fig. 6.26 *A spherical panorama of the interior of the* cella *in the Temple of Bel showing both* thalamoi, *or adytons,* *in the south and the north.* Photo: Gabriele Fangi 2010

(Left) Fig. 6.27 *A spherical panorama of the interior of the cella in the Temple of Bel showing the northern* thalamos. Photo: Gabriele Fangi 2010

(Below) Fig. 6.28 *The northern* thalamos *photographed in 1920–1933.* Library of Congress, Matson Collection

Fig. 6.29 *The northern thalamos.*
Photo: Gabriele Fangi 2010

Fig. 6.30 *The ceiling decorations for
the southern and northern* thalamos
*as illustrated by Wood in 1753.*

Fig. 6.31 *The floral ceiling decoration of the southern* thalamos *photographed in 1920–1933.*
Photo: Library of Congress, Matson Collection

Fig. 6.32 *The floral ceiling decoration of the southern* thalamos. Photo: Gabriele Fangi 2010

Fig. 6.33 *Decorative floral elements in the Temple of Bel.* Photo: Silvana Fangi 2010

Fig. 6.34 *The oriental vine scroll decoration in the Temple of Bel.* Photo: Silvana Fangi 2010

counterpart had Ionian pilasters (Fig. 6.26). In both there were large niches with domed ceilings richly decorated in relief. In the northern *thalamos* ceiling the god Bel appeared in the centre of the dome surrounded with zodiac symbols and supported by four eagles (Fig. 6.30). In the southern *thalamos* ceiling, by contrast, the dome had floral and geometric decorations with a rosette in the centre (Figs. 6.30–6.32).[35] In the *cella* reliefs illustrated the triad of Palmyra in Roman cuirasses, with wailing women, and a camel carrying a ceremonial canopy (Figs. 6.2, 6.4). In addition to floral motifs, oriental vine scrolls formed part of the architectural decoration (Figs. 6.33, 6.34). Assyrian ziggurat-like merlons decorated the top of the roof of the *peristyle*, providing Mesopotamian features for the building (Fig. 6.25). There were stairs to the roof, the staircases being both round and square in plan (see the plan of the *cella* in Fig. 6.16). The roofs were important for celestial and astral cults as in the neighbouring temple of Baalbek in Lebanon. [36]

The building process of the Temple of Bel had been an enormous endeavour that needed to be financed. It was a time of prosperity in the Roman East and the rise of wealth in Palmyra. At the time Palmyra was allegedly a *polis*, a Greek type of city-state, typical in the

Fig. 6.35 *The miniature model of the Temple of Herod with its courtyard in Jerusalem; the temple was built from 20–19* BC. Photo: Kenneth Lönnqvist 1993

Hellenistic world under Roman rule. It governed the surrounding areas. In relation to the Temple of Bel, the Temple of Herod in Jerusalem in the Roman kingdom of Judea is a comparable, nearly contemporaneous, building project comprising some features of the circumambulation plan with a *cella* inside a large courtyard. However, the apocalyptic Ezekiel's temple[37] and the apocalyptic Qumran temple described in the Temple Scroll (11QTemple) found in the caves of Qumran at the Dead Sea are closer to a circumambulation plan comparable with the Temple of Bel.[38] Herod's temple (Fig. 6.35) was erected on a huge artificial and rectangular *temenos* laid out on a mountain; a vast amount of enormous work was required simply for the construction of the platform. Herod's temple also had a *peristyle* in the courtyard, which is mentioned in the Bible[39] and also referred to by the ancient Jewish historian Josephus. That temple was destroyed in AD 70 by the Roman Emperor Titus less than 40 years after the construction of the Temple of Bel.[40] It has been claimed that Palmyrene archers took part in the destruction but there is no confirmation.[41]

## 6.3 The destructions and transformations of the Temple

The Temple of Bel had already been pillaged by the Romans during the conquest of Palmyra by Emperor Aurelian in AD 272. *Historia Augusta, Life of Aurelian* (Part 2, 31, 7–10) refers to Aurelian's letter concerning the Temple of Bel that he calls the Temple of Sun and presents his plan for its restoration:

> Now as to the Temple of the Sun at Palmyra, which has been pillaged by the eagle-bearers of the Third Legion, along with the standard-bearers, the dragon-bearer, and the buglers and trumpeters, I wish it restored to the condition in which it formerly was. You have three hundred pounds of gold from Zenobia's coffers, you have eighteen hundred pounds of silver from the property of the Palmyrenes, and you have the royal jewels. Use all these to embellish the

Fig. 6.36 *Column drums reused in the walls of the* temenos. Photo: Kenneth Lönnqvist 2004, SYGIS

Fig 6.37 *Outer view of the* temenos *walls with windows. Photo: Library of Congress, Maison Bonfils 1867–1899*

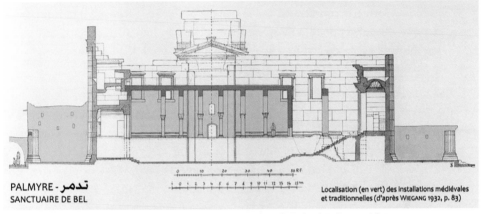

PALMYRE - تـدمـر
SANCTUAIRE DE BEL

Localisation (en vert) des installations médiévales et traditionnelles (d'après WIEGANG 1932, p. 83)

Fig. 6.38 *The green colour shows Medieval and other later additions to the Temple of Bel.* Courtesy by Thibaud Fournet

temple; thus both to me and to the immortal gods you will do a most pleasing service. I will write to the senate and request it to send one of the pontiffs to dedicate the temple.

The surrounding walls of the *temenos* seem to have been partially constructed and reconstructed in many zones from reused architectural members from various times. There are piles of column drums and other structures inserted into the walls (see, for example, Fig. 6.36, 6.37). Pieces of columns have been inserted into the windows of the outer walls as well. The painting of Saint Mary on the walls of the *cella* exemplifies the transformations of the site and is a later development that dates from the Medieval times.[42] But also a *graffito* was incised in the wall hailing Saint Mary by a certain Lazarus.[43] It indicates how the building became reused as a church in the Byzantine era in the 6th to 8th centuries AD. Later on it served as a mosque after the Arab conquests. There were decorated *mihrabs*, niches to mark the direction of Mecca for prayers, that were added to the walls of the *adyton,* which were later moved to the Museum of Damascus[44] (see Fig. 6.38).

What has been not visible for the visitors for decades is the fact that the whole *temenos* courtyard up till 1930 was actually the area of Tadmur village crowded with mudbrick houses (Figs. 6.39–6.43), some dating from Medieval times. Early travellers documented the mud houses in their lithographs.[45] One could then hardly have had a clear view of the major *naos* or *cella* of the temple with its higher gate poking up behind the small square dwellings and narrow streets. Then it must have been difficult to visualise the large *temenos* and feel its spatial impact in the way it was intended.

A traveller, Charles Greenstreet Adison, describes the village or township of Palmyra inside the courtyard of the Temple of Bel in 1838 as follows:

> We entered a vast citadel under a lofty tower, through a gloomy gateway, and found ourselves in the village of Tadmor, which occupies the atient [sic] inclosure, formerly consecrated to the Temple of the Sun. Through a narrow street of dilapidated mud houses, we were conducted into a large court, and received and accomodated by a finely dressed portly Arab lady, adorned with gold coins strung together.[46]

On 21 May 1900 Gertrude Bell wrote: 'I walked out and down and the street of columns into the Temple of the Sun to the town, should I say, for it is nearly all included within its enormous outer walls.' Earlier, on 20 May 1900 she noted: 'The modern town is built inside it and its rows of columns rise out of a mass of mud roofs.'

Both Gertrude Bell and John Garstang had documented the village by photographing it, as did the French. In 1930, during the French Mandate, the village was destroyed, the *temenos* was cleared and the new Palmyra was created outside the ruined area to the north (see Fig. 6.43).

ISIS blew up the Temple of Bel in 2015. The satellites transmitted the images of the destruction to the world (Fig. 6.43). After Palmyra was recaptured by Russian and Syrian forces from ISIS in spring 2016, the Syrian Heritage Project documented the broken pieces of architectural members on the ground in cooperation with ICONEM (a new French start-up using aerial photographs and photogrammetry to help save cultural heritage in danger) and the Syrian Directorate-General of Antiquities and Museums. Remains of bombs and dynamite used in the destruction of the building were detected. The whole site was scanned with the aid of a drone with accuracy of 1 cm, and some blocks were documented with accuracy of 1 mm. A few merlons that once stood on top of the temple had not been completely destroyed. [47]

## 6.4 Crowdsourcing and creating digital restorations of the Temple ruin in 3D

Multiple images provide sources for reconstructing the ruins of the Temple of Bel from various angles. Crowdsourcing is a new way to provide data. Gabriele Fangi and his students have produced a wireframe model of the ruins by applying a screenshot (Fig. 6.44); a CAD (computer-assisted design) model has been created as well (Fig. 6.45). Renderings for creating

Fig. 6.39 *The mudbrick houses of the Tadmor village inside the temenos of the Temple of Bel 1900–1920.* Photo: Library of Congress, the American Colony Collection

Fig. 6.40 *A photograph showing the houses and narrow streets in the precincts of the Temple of Bel in 1900–1920.* Photo: Library of Congress, Matson Collection

(Above) Fig. 6.42 *Restoration work at the Temple of Bel.* Photo: IFPO

(Left) Fig. 6.41 *Mudbrick houses surrounding columns in the temenos in 1900–1920.* Photo: Library of Congress, the American Colony Collection

Fig. 6.43 *The development of the* temenos *and the destruction of the Temple of Bel as seen from a satellite.* Photo: Courtesy of Thibaud Fournet

Fig. 6.44 *A wire frame model constructed by screenshot and crowdsourcing from digital photographs by Gabriele Fangi and Giada Francucci.*

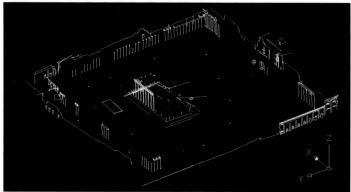

Fig. 6.45 *A CAD model of the Temple of Bel by Gabriele Fangi and Giada Francucci 2016.*

Fig. 6.46 *A solid model of the Temple of Bel by Giada Francucci.*

(Above and left) Figs. 6.47, 6.48 *3D textured models of the Temple of Bel by Wissam Wahbeh and Gabriele Fangi.* Courtesy by Wissam Wahbeh 2016

3D models have also been produced by Giada Francucci (Fig. 6.46), as solid models. Wissam Wahbeh and Gabriele Fangi have produced 3D textured models of the ruins of the Temple of Bel by crowdsourcing from images (Figs. 6.47, 6.48).

For building a digital restoration model of the ruins of the temple as it existed before the destruction by ISIS, ICONEM has also used images that were shot with a drone and on the ground after the destruction.[48] Drawings of structures and texturing with images and augmented reality were added in the creation of a model (Fig. 6.50).

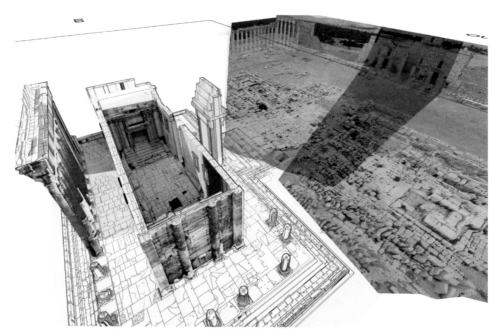

Fig. 6.49 *3D model of the* cella *of the Temple of Bel using augmented reality.* Source: ICONEM

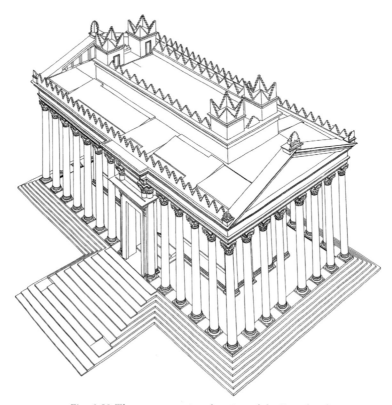

Fig. 6.50 *The reconstruction drawing of the Temple of Bel* cella *produced by Seyrig, Amy, and Will 1975.*

## 6.5 Modelling the past, virtual reconstruction, and immersive experience

Various researchers have drawn and suggested reconstructions of the temple to illustrate how the building might have looked in antiquity. Robert Wood published the plan and his views of the buildings in 1753. Interpretations have since been published by artists as well as Theodor Wiegand (1932), Henri Seyrig, Robert Amy, and Ernest Will (1975) (Fig. 6.50).

In the first chapter we saw the miniature model of the Temple of Bel that existed in the Palmyra Museum (Fig. 1.9). For many years it provided an idea of the ancient structure for visitors. Sometimes ruin romanticism adds more details than the documented evidence justifies. However, the digital era has brought new possibilities of displaying the original texture of surfaces using digital images.

As mentioned in Chapter 1, the New Palmyra Project was established by a Palestinian engineer before the Syrian Civil War broke out in 2011. The purpose was to create virtual models of Palmyra showing how the structures might have looked in ancient times. As mentioned, this is an Open Source Project that gathers images from the general public.[49]

The Virtual Palmyra Project led by Ahmet Denker has based its interpretations of the Temple of Bel on information from ancient sources, lithographs, and architectural studies by Wiegand, Seyrig, Amy, and Will (Figs. 6.51–6.53). The project has added look-alike landscapes to the reconstructions that offer a means to experience the past (Fig. 6.54). Also, immersive events have been provided for these reconstructions of the ancient building and spaces using aviation and video technologies. This travelling to the virtual past is a vivid experience, especially feeling the spaces such as the colonnades and porticos surrounding

Fig. 6.51 *The virtual model of the façade of the Temple of Bel* cella *constructed by Ahmet Denker 2016.*

Fig. 6.52 *The flank of the* cella *of the Temple of Bel and surrounding porticos in the courtyard.*
Constructed by Ahmet Denker 2016

Fig. 6.53 *A virtual reconstruction of the portico surrounding the cella of the Temple of Bel.*
Constructed by Ahmet Denker 2016

Fig. 6.54 *A virtual reconstruction of the Temple of Bel with its imaginary landscape.*
Constructed by Ahmet Denker 2016

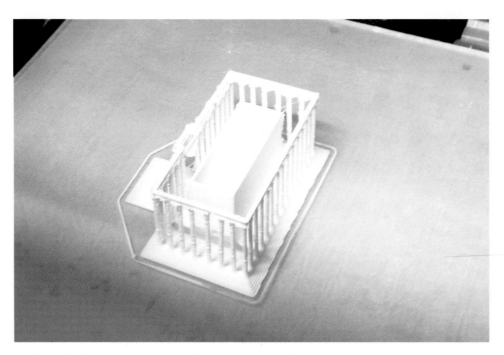

Fig. 6.55 *3D printed miniature replica of the Temple of Bel produced by Ahmet Denker 2016.*

the temple courtyard. Moving through the spaces provides extra value to the encounter. Music, if used, adds feelings to the movement, and the virtual visit can heal some of the trauma felt after the destruction. In addition, Denker has used 3D printing technology to make miniature replicas of the Temple of Bel (Fig. 6.55), including the surrounding courtyard. These can be used in exhibitions and serve the audiences in the way the old miniature model enlightened the visitors in the Palmyra Museum.

## Endnotes

1   See Teixidor, J. (1979) *The Pantheon of Palmyra*. Leiden: Brill; Teixidor, J. (1980) Cultes tribaux et religion civique à Palmyre, in *Revue d'histoire des religions*, Vol. 197, pp. 277–287.

2   Lipinski, E. (1983) The God 'Arqû-Rashap in the Samallian Hadad Inscriptions, in *Arameans, Aramaic and the Aramaic Literary Tradition*, ed. by Sokoloff, M., Bar-Ilan Studies in Near Eastern Languages and Cultures, Ramat Gan: Bar-Ilan University, pp. 15–21.

3   Teixidor, J. (1979) *The Pantheon of Palmyra*, Leiden: Brill, pp. 69–70.

4   Dirven, L. (2010) Religious Frontiers in the Syrian-Mesopotamian Desert, in *Frontiers in the Roman World, Proceedings of the Ninth Workshop of the International Network Impact of Empire (Durham 16–19 April 2009)*, ed. by Hekster, O, & Kaizer, T., Leiden: Brill, pp. 157–174.

5   Al-As'ad, Kh. and Gawlikowski, M. (1997) *The Inscriptions in the Museum of Palmyra, A Catalogue*, Palmyra and Warsaw: The Commitee for Scientific Research, Warsaw, Poland, pp. 20–21; Al-As'ad and Gawlikowski (see above) hold that it is possible that the temple, even if inaugurated in AD 32, was actually finished later in the 1st century AD; see also Teixidor, J. (1979) *The Pantheon of Palmyra*, Leiden: Brill, pp. 1–16.

6   Dirven, L. (1999) *The Palmyrenes at Dura Europos: a Study of Religious Interaction in Roman Syria*. Leiden: Brill, pp. 43–44.

7   See, for example, Teixidor, J. (1979) *The Pantheon of Palmyra*, Leiden: Brill, pp. 29–52.

8   See Teixidor, J. (1979) *The Pantheon of Palmyra*, Leiden: Brill, pp. 106–109.

9   For example, I Kings 21–22.

10  See Ragette, P. (1980) *Baalbek*. Park Ridge: Noyes Press.

11  See Halsberghe, G. (1972) *The Cult of Sol Invictus*. Leiden: Brill.

12  Dirven, L. (1999) *The Palmyrenes at Dura Europos: a Study of Religious Interaction in Roman Syria*. Leiden: Brill.

13  Hauser, S. (2007) Temple für den palmyrischen Bel, in *Getrennte Wege? Kommunikation, Raum und Wahrnehmung in der Alten Welt*, Herausgegeben von Rollinger, R., Luther, A. and Wiesehöfer, J. unter Mitarbeit von Gufler, B., Frankfurt am Main: Antike Verlag, pp. 228–256, see also Tepstra, T.T. (2016) The Palmyrene Temple in Rome and Palmyra's Trade with West, in *Palmyrena: City, Hinterland and Caravan Trade between Orient and Occident*, ed. by Meyer, J.C., Seland, E.H. and Anfinset, N., Oxford: Archaeopress, pp. 39–48.

14  Bryan, D.B. (1973) *Texts Relating to the Marzeah: A Study of an Ancient Semitic Institution*, UMI dissertations, Ann Arbor, Michigan, USA, p. 2; see *marzeah* in Ugarit KTU 1.114; 3.9; see also in the Bible: Jeremiah 16:5; Amos 6:7.

15  As mentioned in Ch. 5, the SYGIS project led by the present principal author found one *tessera* in a Roman post of Qseybe under the area of Palmyra beside pottery with inscribed Aramean signs in the territory zone of Palmyra.

16  Vitruvius, *The Ten Books on Architecture*, Translated by Morris Hicky Morgan, New York: Dover Publications (1960), Chap. I, 1, 3.

17  Browning, I. (1979) *Palmyra*, Park Ridge: Noyes Press, pp. 102–108.

18  See Schulz, B. (1932) Das grosse Hauptheiligtum des Bel, in Wiegand, T. (1932) *Palmyra – Ergebnisse der Expeditionen von 1902 und 1917*, Vol. I, II, Textband, Tafeln, Archeologisches Institut des Deutschen Reiches Abteilung Istanbul, Berlin: Verlag von Heinrich Keller, p. 127–150.

19  Al-As'ad, Kh. and Gawlikowski, M. (1997) *The Inscriptions in the Museum of Palmyra, A Catalogue*, Palmyra and Warsaw: The Committee for Scientific Research, Warsaw, Poland, pp. 20–21; Browning. I. (1979) *Palmyra*, Park Ridge: Noyes Press, p. 107.

20  See Schulz, B. (1932) Das grosse Hauptheiligtum des Bel, in Wiegand, T. (1932) *Palmyra – Ergebnisse der Expeditionen von 1902 und 1917*, Vol. I, II, Textband, Tafeln, Archeologisches Institut des Deutschen Reiches Abteilung Istanbul, Berlin: Verlag von Heinrich Keller, pp. 127–150.

21 Seyrig, H., Amy, R. and Will, E. (1975) *Le Temple de Bel a Palmyre*, Vol. I, II. Album, Texte et Planches. Institut Français d'Archéologie de Beyrouth. BAH. T. 83. Paris: Librairie orientaliste Paul Geuthner.

22 Hammad, H. (2006) *Le Sanctuare de Bel à Tadmor-Palmyre*, Paris: Geuthner, pp. 255–256.

23 Jericho presently in the Palestinian territories of Israel northwest of the Dead Sea has provided a rectangular temple dating from the Neolithic period; see Kenyon, K. and Holland, T.A. (1981) *Excavations at Jericho*, Vol. III, The Architecture and Stratigraphy of the Tell, London: the British School of Archaeology in Jerusalem, pp. 306–307.

24 M. Saghieh holds that the Megaron form emerged in Anatolia in the Chalcolithic Age: see Saghieh, M. (1983) *Byblos in the Third Millennium B.C.: A Reconstruction of the Stratigraphy and a Study of the Cultural Connections*, Warminster, Wiltshire: Aris & Phillips, p. 124.

25 See, for example, Cooper, L. (2006) *Early Urbanism on the Syrian Euphrates*, London & New York: Routledge, pp. 148–161.

26 Matthiae, P. (1979) Ebla in the Period of Amorite Dynasties and the Dynasty of Akkad, in *MANE* (Sources and Monographs on the Ancient Near East, Malibu), Vol. 1, Fasc., 6, p. 20, Fig. 6; see also Matthiae, P. (1975) Unité et développement du Temple dans la Syrie du Bronze Moyen, in *Le Temple et Le Culte*, Compte rendu de la vingtième Rencôntre Assyriologique Internationale, ed. by Donzel, E., Donceel-Voûte, P., Kampan, A.A., and Mellink, M.J., Istanbul: Nederlands Historisch Instituut te Istanbul, pp. 43–72.

27 II Chronicles 2.

28 Lönnqvist, M. (2000) *Between Nomadism and Sedentism: Amorites from the Perspective of Contextual Archaeology*, Helsinki: Juutiprint, pp. 244–246.

29 See Anadol, S. (2008) Palmyra – Identity Expressed through Architecture and Art, in *Jebel Bishri in Context: Introduction to the Archaeological Studies and the Neighbourhood of Jebel Bishri in Central Syria*, ed. by Lönnqvist, M., BAR International Series 1817, Oxford: Archaeopress, pp. 59–72, 68.

30 Hammad, H. (2006) *Le Sanctuare de Bel à Tadmor-Palmyre*, Paris: Geuthner, p. 252.

31 Browning, I. (1979) *Palmyra*, London: Chatto & Windus, pp. 118–119.

32 Browning, I. (1979) *Palmyra*, London: Chatto & Windus, pp. 118–119.

33 Browning, I. (1979) *Palmyra*. London: Chatto & Windus; Hammad, M. (2006) *Le Sanctuare de Bel à Tadmor-Palmyre*, Paris: Geuthner, p. 238.

34 Seyrig, H., Amy, R. and Will, E. (1975) *Le Temple de Bel a Palmyre*, Vol. I, II, Album, Texte et Planches, Institut Français d'Archéologie de Beyrouth, BAH, T. 83, Paris: Librairie orientaliste Paul Geuthner, p. 170.

35 Hammad, H. (2006) *Le Sanctuare de Bel à Tadmor-Palmyre*, Paris: Geuthner, pp. 241–243; Hammad, M. and Fabbri, P. (2016) *Bel/Palmira: Hommage*, Paris: Geuthner/Guaraldi, p. 81.

36 See Amy, R. (1950) Temples à Escaliers, in *Syria*, Vol. XXVII, pp. 82–136.

37 Hesekiel 41, 42.

38 See Lönnqvist, M. and Lönnqvist, K. (2002) *Archaeology of the Hidden Qumran: The New Paradigm*, Helsinki: Helsinki University Press, pp. 234–235.

39 Josephus, *Bellum Judaicum = The Jewish War*, Book VI; see also Smallwood, E. M. (1976) *The Jews under Roman Rule: from Pompey to Diocletian, a Study of Political Relations*, Leiden: Brill; see also a more recent study by Goodman, M. (2007) *Rome and Jerusalem: the Clash of Ancient Civilizations*. London: Penguin.

40 Smallwood, E.M. (1976) *The Jews under Roman Rule: from Pompey to Diocletian, a Study of Political Relations*, Leiden: Brill, pp. 531–532, note 19; see also Starcky and Gawlikowski citing The Babylonian Talmud that was expecting the fall of Palmyra claiming that 80, 000 archers from Palmyra had participated in the destruction of the First Temple and 8, 000 in the Second Temple; see also Starcky, J. and Gawlikowski, M. (1985) *Palmyre*, Paris: Libraire Amérique et d'Orient, J. Maissonneuve soc., p. 40.

41 Hammad, M. and Fabbri, P. (2016) *Bel/Palmira: Hommage*, Paris: Geuthner/Guaraldi, p. xxvi.

42 See, for example, As'ad, Kh. and Yon, J.-B. (2001) *Inscriptions de Palmyre: Promenades épigraphiques dans la ville antique de Palmyre*, Guides achéologiques de l'IFAPO, Beyrouth-Damas-Amman: Institut français d'archéologie du Proche-Orient, p. 95.

43 Hammad, M. and Fabbri, P. (2016) *Bel/Palmira: Hommage*, Paris: Geuthner/Guaraldi, p. xxvi.

44 See, for example, As'ad, Kh. and Yon, J.-B. (2001) *Inscriptions de Palmyre: Promenades épigraphiques dans la ville antique de Palmyre*, Guides achéologiques de l'IFAPO, Beyrouth-Damas-Amman: Institut français d'archéologie du Proche-Orient, p. 98.

45 Greenstreet Addison, C. (1838) *Damascus and Palmyra: a journey to the East. with a sketch of the State and Prospects of Syria under Ibrahim Pasha*, Vol. 2, London: Richard Bentley, p. 252.

46 https://sketchfab.com/iconem, the website entered 26th August, 2016.

47 https://sketchfab.com/iconem, the website entered 26th August, 2016.

48 See, for example, Busta, H. (2015) An Open-Source Project to Rebuild Palmyra, in *Architect Magazine*, October 23, 2015.

# 7

# THE TRIUMPHAL ARCH AND
# THE GRAND COLONNADE

## 7.1 The Triumphal Arch, the symbol of Palmyra

The imaginary scene in the painting *James Dawkins and Robert Wood Discovering the Ruins of Palmyra* by Gavin Hamilton (1758) points only to one central monument in Palmyra. It is the Triumphal or Monumental Arch – a great gateway to the city and its history (see Ch. 1). Drawings and paintings since Robert Wood's phenomenal work (Fig. 7.2) published

Fig. 7.1 *A spherical panorama of the Triumphal Arch of Palmyra.* Photo: Gabriele Fangi 2010

in 1753 show the arch as the focal point of the caravan city. Old photographs illustrated it in sepia or black and white (Figs. 7.3–7.5). Early visitors and tourists were photographed in front of it (Fig. 7.6). The early 20th century colour photographs also documented the arch (Fig. 7.7).

The arch was one of the monuments that ISIS destroyed in 2015, and one that was not a religious building like the others that were blown up. However, like the Temple of Bel, the arch was a major landmark of Palmyra and its symbolism of power was visible. It stood for the identity of the caravan city. The reason it was destroyed remains somewhat a mystery, but it may be assumed that its major importance in relation to the place attracted the jihadists to make it a propagandist target to show to the western world that they had the power to erase such monuments from history. Also, the arch was a Roman enterprise that demonstrated its western mastery over the Oriental city and given its significance as a triumphal monument its destruction can be seen as a subversive act.

(Left) Fig. 7.2 *A drawing of the Triumphal Arch published by Robert Wood in 1753.*

(Below) Fig. 7.3 *The Triumphal Arch photographed in 1867–1899.* Photo: Library of Congress, Maison Bonfils, Beirut, Lebanon

Fig. 7.4 *The Triumphal Arch photographed in 1900–1920*. Photo: Library of Congress

Fig. 7.5 *The Triumphal Arch photographed in 1920–1933*. Photo: Library of Congress, Matson Collection

Fig. 7.6 *An automobile and visitors in Palmyra photographed in front of the Triumphal Arch.*
Photo: Library of Congress

Fig. 7.7 *The Triumphal Arch photographed in 1932–1951. Bedouins and a
French official riding on camels through the arch.* Photo: Library of Congress

The Triumphal Arch was a passage, a gateway, which stood at the southeastern end of the Grand Colonnade, the main street leading to smaller sanctuaries and civic sites along the street, to the markets of Palmyra or taking one in the opposite direction – the southeast – towards the Temple of Bel. Both the arch and the colonnade have been popular cover images of books on Syria and the Near East, characterising the Roman Near East as a whole. The image of the arch has been depicted on modern Syrian coins as the symbolic monument of Syria itself.

The Triumphal Arch was built from limestone during the reign of Emperor Septimius Severus (ruled AD 193–211) and is dated to the late 2nd century AD. Its ground plan was designed as a V to provide an optical illusion by turning the orientation of the arch by 30° to the street in order that visibility was provided from the Grand

Fig. 7.8 *The Arch of Septimius Severus at the Forum in Rome in Italy.* Photo: Kenneth Lönnqvist 2008

Fig. 7.9 *Architectural details such as ornamentation in the Triumphal Arch, photographed 1920–1933.* Photo: Library of Congress, the American Colony collection

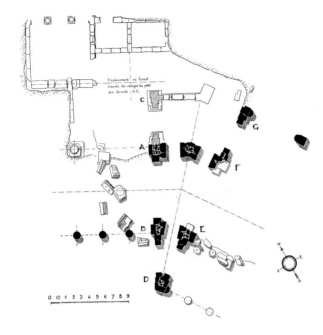

(Left) Fig. 7.10 *Scale plan of the site of the Triumphal Arch, published by R. Amy in 1933.*

(Below) Fig. 7.11 *Architectural scale drawings used in the restoration of the Triumphal Arch provided by R. Amy in 1933.*

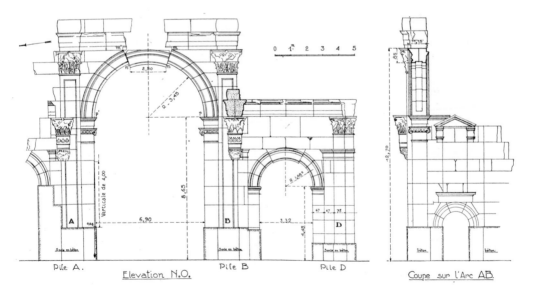

Colonnade towards the Temple of Bel (Fig. 7.10) through the arch, emphasising the temple. The floral and leaf decorations with egg and dart linings are magnificent and clearly Palmyrene (Fig. 7.9).[1] The inner ceiling follows the pattern of Roman arches, for example that of Semptimius Severus's triumphal arch (Fig. 7.8) in *Forum Romanum* in Rome with geometric patterns inserted with flowers. The Septimius Severus arch in Rome, being so well preserved, can provide us a glimpse of how the original form of a Roman triumphal arch was conceived.

Figs. 7.12–7.14 *3D models of parts of the Triumphal Arch constructed by Gabriele Fangi and Clara Forino 2016.*

Fig. 7.15 *An imaginary virtual reconstruction of the Triumphal Arch constructed by Ahmet Denker 2016.*

The monument was restored by Robert Amy in 1930. He also conducted an excavation, finding some of its fallen parts, such as capitals for the pilasters, for its *anastylosis* (the raising of the architectural members to their original places). The central great arch was flanked by smaller ones on both sides. The larger (AB) and smaller (BD) arches faced northwest and the small arches (DE and FG) faced southeast (Figs. 7.10, 7.11). The large arch needed urgent restoration; some fractures had been caused by an earthquake.[2]

In relation to the Triumphal Arch, there was a Greek inscription on the base of a statue declaring victory over the Persians achieved by Septimius Herod (apparently the same Herod as Hairan in Aramaic and son of Zenobia) crossing the Orontes River. It mentions Aurelius Septimius Worod and some other dignitaries such as Iulius Aurelius, a procurator, and both bear the title of *strategos*. The name of the latter is only partly preserved. The inscription dates c. AD 268–269 (IGRR III, 1032).

## 7.2 Digitally imaging, reconstructing, and printing gateways

Digital imaging of arches and gateways has provided ways to reconstruct the ruin of the Triumphal Arch digitally in 3D (Figs. 7.12–7.14) and virtually recreate the possible spatial scene in antiquity (Fig. 7.15).

The Institute for Digital Archaeology (IDA) has built a replica of the Triumphal Arch through crowdsourcing existing digital images and applied 3D printers for erecting it in marble. The central arch was printed and carved in Carrara in Italy and erected in London's Trafalgar Square in the spring 2016 and in New York in the autumn 2016. The work was a statement and a gesture to show that what the jihadists have destroyed will be rebuilt. Other arches for remembering and overcoming the trauma have been designed. For example, artists, computer specialists, and students from the Massachusetts Institute of Technology (MIT) designed a

Figs. 7.16, 7.17 *The 3D printed replica of the central arch from the Triumphal Arch partly produced at Carrara in Italy.* Courtesy: The Institute for Digital Archaeology

(Above) Fig. 7.18 *The erection of the 3D printed replica in Trafalgar's Square in London in 2016.* Courtesy: The Institute for Digital Archaeology

(Right) Fig. 7.19 *The Great Gateway to Baalbek by David Roberts.*

mobile arch of Palmyra to take a stand for peace. This mobile arch was decorated with numerous suspended plastic squares which, like wind chimes made sounds, in the breeze.[3]

3D printing technology opens new opportunities. 3D replicas can help as a substitute, and one can build museum exhibitions using prints. Seeing a replica provides kinetic ways to visually comprehend the monument that we are not able to visit. Touching the prints allows us to understand the feeling of the shapes and surfaces. Moulding and casting have been carried out since antiquity, and casts have been made of various materials. Ancient coroplasts (makers of terracotta figures) and sculptors showed that clay was valuable material for modelling.[4] It is clear that the ultimate reconstruction of the Triumphal Arch in Palmyra needs the original pieces that were left from the destruction. Their documentation and collection are essential undertakings, if and when the restoration is to take place at the site of Palmyra.

Interestingly, the portal to the *cella* of the Temple of Bel was the only surviving structure of the temple after it was blown up by the jihadists, standing lonely in the midst of the rubble. Betty Ratcliffe's delightful miniature from the late 18th century (see Ch. 1) seems to show the portal of Bel's temple, or alternatively, it can also reflect various ruins including Baalbek and its gateway (Fig 7.19). The great gateway of Baalbek has also been the subject of artworks in the past. The gateways of Palmyra – the Triumphal Arch and the portal of the *cella* – and their illustration have also become the subject of rehabilitation of cultural memory that was carried out among the children of Syria during the civil war by the Institute for Digital Archaeology.

The walls of Palmyra had numerous gates for the traffic, such as the previously mentioned Damascus Gate and the Gate of Dura. In addition to the Triumphal Arch, Palmyra had several other arches, some in the Grand Colonnade providing access to transverse streets (see Figs. 7.25, 7.26, 7.32). The *Tetrapylons* (Figs. 7.33–7.35), both in the Grand Colonnade as well as in the Camp of Diocletian situated at the crossroads, also might have served as gate structures. In the Camp of Diocletian in the western part of Palmyra stood the *Porta Praetoria* that led to the military *castrum* and the Grand Gate that led to the Forum.[5]

Fig. 7.20 *Lintel of a gateway in Palmyra photographed in 1920–1933*. Library of Congress, American Colony collection

(Above) Fig. 7.21 *News in the* Guardian *newspaper about the destruction to the Temple of Bel in Palmyra in 2015. Only the portal of the temple remained standing after demolition of the temple by ISIS.* Photo: Getty Images

(Left) Fig.7.22 *The portal to the Temple of Bel.* Photo: Silvana Fangi 2010

(Left) Figs. 7.23, (below) 7.24 *The panorama of the Grand Colonnade and the colonnade seen from the air.* Photos: SYGIS 2000 and Gabriele Fangi 2010

## 7.3 The Grand Colonnade

The Grand Colonnade was a paved street. It is a magnificent architectural endeavour of yellow limestone columns of the Corinthian order running roughly from southeast to northwest. The street stretches c. 1,200 metres. It was full of fallen architectural members when earlier visitors approached the city and its monuments. Apart from limestone, some columns are of granite, such as the *Tetrapylon* erected of monolithic columns carved from the red granite of Aswan, and the portico of the Diocletian Baths inserted in to the Grand Colonnade. The *Tetrapylon* is at the beginning of a smaller diagonal transversal street leading to a piazza and the Damascus Gate in the southwest.[6]

Aerial photographs (Fig. 7.23) provide a clear plan of the street, and panoramic images (Fig. 7.24) allow one to visualise the street from the south. The course of columns provides for a spectator a visual rhythm emphasised by the brackets (see Figs. 7.25–7.30). This is a great work of ancient architecture in which harmony is achieved in mathematical planning. The street created the view from the *Tetrapylon* to the Triumphal Arch and in turn from the Triumphal Arch towards the *Tetrapylon*

Fig. 7.25 *The Grand Colonnade photographed in 1867*. Photo: Library of Congress, Maison Bonfils collection

Fig. 7.26 *The western colonnade photographed in 1900–1920.*
Photo: Library of Congress, American Colony collection

Fig. 7.27 *Visitors at the Grand Colonnade*. Photo: Library of Congress

Fig. 7.28 *The Grand Colonnade*. Photo: Gullög Nordquist 2004, SYGIS

Fig. 7.29 *The Grand Colonnade.* Photo: Gullög Nordquist 2004, SYGIS

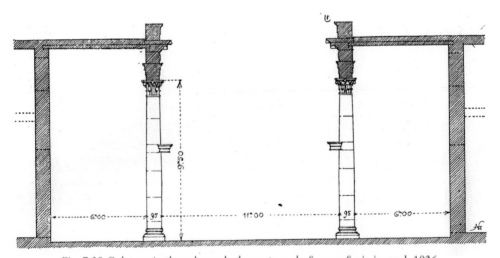

Fig. 7.30 *Columns in the colonnade drawn to scale.* Source: Syria journal, 1926

(Figs. 7.28, 7.29, and 7.33). The column brackets once supported statues of various dignitaries of Palmyra, and the faces of the columns preserve the history of the city in inscriptions that commemorate the deeds and contributions of the famous members of the society.

In the northeastern section of the colonnade, there is a Greek inscription that could refer to the construction of the colonnade and dates to AD 158: 'The Council and the People (honor) Zebîdâ Taîmê, son of Moqîmû, son of Garbâ, the constructor...' In the Grand Colonnade there is another important inscription referring again to Septimius Worod. His name appears to be

Fig. 7.31 *A virtual reconstruction showing the visual effects of a colonnade like the Grand Colonnade in antiquity. The reconstruction does not include buildings and transversal streets.* Construction by Ahmet Denker 2016

Fig. 7.32 *An arch leading to a transversal street deviating to the south from the Grand Colonnade.* Photo: Kenneth Lönnqvist 2004, SYGIS

Iranian. He was a central public figure in the commerce of Palmyra, the chief of the markets, sometime during the rule of Odenathus and Zenobia. He was a *strategos,* a person of the *Agora*, and the chief of the priests in the Temple of Bel (CIS 3942).

The Grand Colonnade provided access to several public buildings and spaces along it. The earliest section of the street, the so-called upper colonnade running from the northwest, apparently ran from the Funerary Temple towards the *Tetrapylon* that created a kind of gate structure of its own (Figs. 7.34, 7.35). This magnificent monument was partly unearthed, studied, and restored in the 1960s. It consisted of 16 red granite Corinthian columns that were erected in the restoration. The original height of the columns was calculated from examining fragments and by comparison with other buildings in Palmyra.[7] The structure was demolished by ISIS at the beginning of 2017.

146

Fig. 7.33 *The view from the Grand Colonnade towards the Tetrapylon.*
Photo: Kenneth Lönnqvist 2004, SYGIS

Fig. 7.34 *The* Tetrapylon *of the Grand Colonnade.* A stitched photo: Gabriele Fangi 2010

Fig. 7.35 *A virtual model of the* Tetrapylon *as it could have looked in antiquity.* Constructed by Ahmet Denker 2016

The neighbourhood of the *Tetrapylon* was a special place for the honorific inscriptions referring to rulers but the site also includes other honorific inscriptions that illustrate the careers of notable Palmyrenes. With a statue and a bilingual inscription in both Greek and Aramaic, the Council and the People honour Iulius Aurelius Zabdilah-Zènobios, who had been a *strategos* of the colony during the rule of Emperor Severus Alexander in AD 242–243 (IGRR III, 1033; PAT 0278). Another important Aramaic inscription is associated, apparently posthumously, with the statue of Odenathus as the king and ruler of the whole East in AD 271 (PAT 0292). Another that mentions Zenobia and her statue in the same year is bilingual with Greek and Aramaic texts: 'Statue of Septimia Zenobia (Batzabbaî in Aramaic), a very illustrious and pious queen erected to their matron by generals of the troops Septimius Zabdâ and Septimius Zabbaî in AD 271' (IGRR III, 1030; PAT 0293).

## Endnotes

1   See Gabriel, A. (1926) Recherches archéologiques a Palmyre, in *Syria*, Vol. VII, pp. 71–92, especially p. 81, 82; Amy, R. (1933) Prémier restaurations de l'Arc monumental de Palmyre, in *Syria*, Vol. 14, pp. 396–411; Browning, I. (1979) *Palmyra*, London: Chatto & Windus, pp. 87–89.

2   Amy, R. (1933) Prémier restaurations de l'Arc monumental de Palmyre, in *Syria*, Vol. 14, pp. 396–411.

3   Presentations at *The World Heritage Strategy Forum*, 9–11 September, 2016, Harvard University.

4   See Silver, M. (2016) Conservation Techniques in Cultural Heritage, in *3D Recording, Documentation and Management of Cultural Heritage*, ed. by Stylianidis, E. and Remondino, F., Caithness, Scotland: Whittles Publishing, pp. 15–105, especially pp. 18–19, and p. 77.

5   Michalowski, K. (1961) *Palmyre: Fouilles Polonaises 1962*, Warsaw: Editions scientifiques polonaises, pp. 9–38; Michalowski, K. (1966) *Palmyre: Fouilles Polonaises 1963 and 1964*. Warsaw: Editions scientifiques polonaises; see also Browning, I. (1979) *Palmyra*, Park Ridge, New Jersey: Noyes Press, p. 185.

6   A. Gabriel considered the Triumphal Arch with the colonnaded street to represent primarily Hellenistic style; see Gabriel, A. (1926) Recherches archéologiques a Palmyre, in *Syria*, Vol. VII, pp. 71–92, especially p. 81, 82; see Bounni, A. and Al-As'ad, Kh. (1997) *Palmyra: History, Monuments and Museum*, Damascus, p. 66, 80; see also a recent study on colonnaded streets with references to the earlier works of R. Burns and K. Butcher by Bühring, C. (2016) The Stage of Palmyra: Colonnaded Streets, Spaces for Communication and Activities in the Eastern Roman Empire, in *Palmyrena: City, Hinterland and Caravan Trade between Orient and Occident*, ed. by Meyer, J.C., Seland, E.H. and Anfinset, N., Oxford: Archaeopress, pp. 59–75.

7   Ostrasz, A. (1966) Etudes sur la restauration du Grande Tetrapyle, in *Studie Palmyrenskie*, Vol. I, pp. 46–58.

# 8

# SACRED AND PUBLIC SPACES ALONG THE GRAND COLONNADE

## 8.1 The Temple of Nabu

From the Triumphal Arch to the left and from the Grand Colonnade to the south excavations took place at a temple dedicated to the Mesopotamian god Nabu, or Nebo, the son of Marduk Bel, in 1963–1970. Additional studies took place in 1977–1981 under the direction of Adnan Bounni and Nassib Saliby from the Directorate-General for Antiquities and Museums of Syria.[1] *Tesserae* associate Nabu with the Greek god Apollo.[2] Earlier, during the studies of Theodor Wiegand, published in 1932, the temple was simply referred to as 'the Corinthian temple' before the dedicatee god of the temple was identified[3] (See Figs. 8.1 and 3.15).

The first, second, and third phases of the building of the temple date from the first, second, and third centuries AD, respectively. The *cella* of the temple covers 20.6 metres × 9.15 metres in plan and was surrounded by a Corinthian porticoed courtyard. The temple was oriented to the south. In some ways the temple resembles the Temple of Bel, with a surrounding courtyard, staircase towers, and Assyrian merlons as decorative elements on the *cella*. In the north of the *cella*, there was a *thalamos* or *adyton* as in the Temple of Bel.[4]

Fig. 8.1 *The ruins of the Temple of Nabu (Nebo) photographed from above in 1910.*
Photo: IFPO

(Left) Fig. 8.2 *The reconstruction of the* propylaea, *the gate building, to the Temple of Nabu published by Wiegand 1932.*

(Below) Fig. 8.3 *The plan of the Temple of Nabu by Wiegand 1932.*

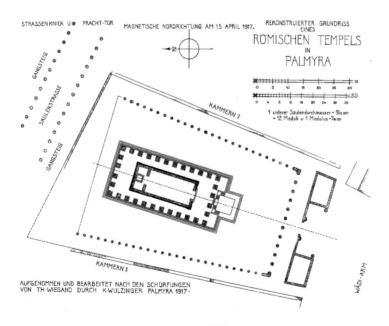

(Right) Fig. 8.4
*The reconstruction
of the pronaos
of the Temple of
Nabu by Wiegand
1932.*

(Below) Fig. 8.5
*The ruins of the
Temble of Nabu.*
Photo: Kenneth
Lönnqvist 2004,
SYGIS

The statue of the sun-god with two eagles was situated in a cultic niche. Another statue of an enthroned goddess, apparently Tykhe guarding the city of Palmyra, and dating from the 1st century AD, is among the most important found at the site.[5]

As in the Temple of Bel, here Greco-Roman architectural orders are fused with Oriental elements. There was, according to Wiegand, a small *propylaeum* building through which one entered the colonnaded courtyard (Fig. 8.2). In the Temple of Nabu, the courtyard surrounding the *cella* is much smaller than in the Temple of Bel and somewhat trapezoidal in shape, being 44 metres wide on the southern side, 85 metres on the eastern, 87 metres on the western, and 60 metres on the northern (see Fig. 8.3). Three porticos have pseudo-Doric columns, although some identify them as Tuscanian columns (Fig. 8.5), and the northern one was reused for building shops along the Grand Colonnade. Many inscriptions have been revealed in the excavations of the temple. Wealthy families of Palmyra funded the temple including the family Elahbel that had an imposing tower tomb dating from AD 103 in the Valley of the Tombs.[6]

## 8.2 The Temple of Baalshamin

The Grand Colonnade provides a route to the north to the Temple of Baalshamin (Figs. 8.5–8.10, 8.12), which had been well excavated and studied before ISIS blew the building up in August 2015, one week after the execution of Khaled Al As'ad, the retired director of the Palmyra Museum.[7] Robert Wood published the drawing of the *cella* of the temple for his great opus of the ruins in 1753 (Fig. 8.6, 8.10)[8] as had Louis François Cassas in his picturesque travels in 1798–1804.[9] Wiegand studied the site with his team in 1902 and 1917 publishing fine drawings of the temple in 1932[10] (Figs. 8.7–8.9). At the request of UNESCO a Swiss mission was employed by the Syrians and led by Paul Collart and Jacques Vicari, who excavated and thoroughly studied the site in 1954–1956[11] (see map, Fig. 3.15).

The temple was originally surrounded by porticos (see a virtual reconstruction Fig. 8.14) like the Temple of Bel and the Temple of Nabu. A *cella* had a *pronaos* with six Corinthian columns, and Corinthian pilasters surround the walls of the building. The plan of the *cella* covered 16.10 metres × 9.91 metres with total outer area covering 17.23 metres × 10.79 metres[12] (Fig. 8.7). Here again inside the *cella* there existed an *adyton* or *thalamos*, with a cultic niche. The capitals of the western portico display Egyptian features with lotus flowers reminiscent of the Ptolemaic capitals. The Egyptian influence from the Greco-Roman period adds Hellenistic nuances by fusing Oriental touches to the Greek architectural order.[13]

There are old photographs that illustrate the sanctuary in the 19th

Fig. 8.6 *Drawing of the Temple of Baalshamin in Palmyra published by Wood in 1753.*

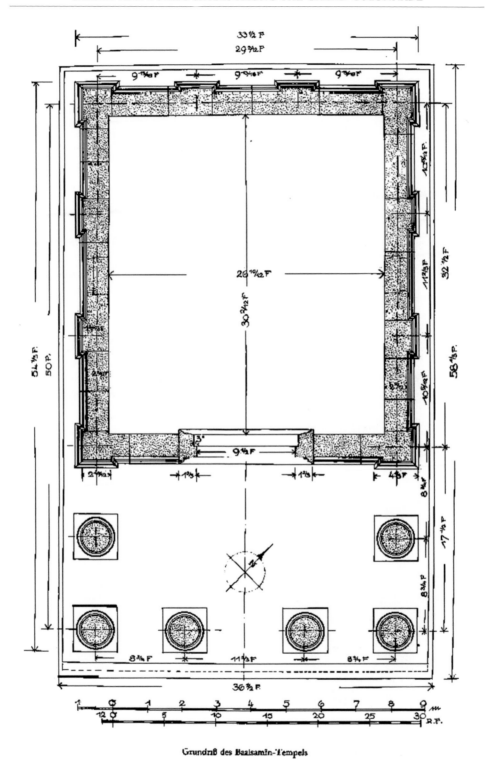

Grundriß des Baalsamin-Tempels

Fig. 8.7 *The plan of the Temple of Baalshamin published by Wiegand 1932.*

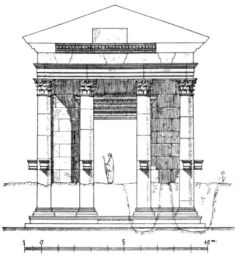

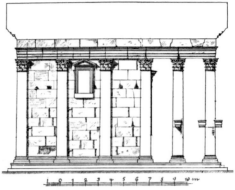

(Left) Fig. 8.8 *Partly reconstructed front of the Temple of Baalshamin by Wiegand 1932.*

(Above) Fig. 8.9 *Partly reconstructed flank of the Temple of Baalshamin by Wiegand 1932.*

Fig. 8.10 *Suggested reconstruction of the Temple of Baalshamin by Wood 1783, the measurements are in feet.*

century and early 20th century. In 1900 Gertrude Bell had pitched her tent in front of the temple with her attendants, horses and mules documenting her stay by photographing the site; her photographs are now in the collection of the University of Newcastle. John Garstang photographed the sanctuary in the 1920s. From the same period, there are photographs in the Matson Collection in the Library of Congress (Figs. 8.11, 8.12). In addition, early photographs of the temple are to be found in the Getty collection and the collections of IFPO as well.

There is an inscription stating that two columns in the temple were dedicated to Baalshamin by 'Attai and Šebhai in AD 23. A dedication from AD 52 has been found on another column.[14] An honorific inscription by Males Agrippa, son of Yarhai, secretary of the city, securely dates the building that stood in the forum until 2015 to AD 130/131.[15] The date coincides with Emperor Hadrian's visit to Palmyra.

Fig. 8.11 *The Temple of Baalshamin photographed in 1900–1920. Some parts such as columns are still buried.* Library of Congress, Matson Collection

Fig. 8.12 *The Temple of Baalshamin photographed in 1900–1920.* Library of Congress, Matson Collection

Fig. 8.13 *The Temple of Baalshamin as fully revealed after the excavations.* Photo: Gullög Nordquist 2004, SYGIS

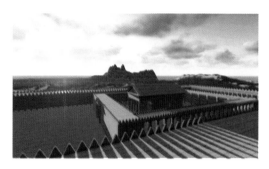

Fig. 8.14 *Virtual reconstruction of the Temple of Baalshamin showing how the building could have looked like in antiquity.* Constructed by Ahmet Denker 2016

The temple that had been built of limestone had apparently been erected by the tribe of Bene Maʾazîn, the Sons of Maʾazîn. The building was an initiative of the same family as the Temple of Allat[16] that is situated in the Camp of Diocletian area in the western corner of the Roman city. The tribe incorporates in its name the word *bene*, the sons, like the previously mentioned tribes living in the region during the Bronze and Iron Ages. Baalshamin was clearly a West Semitic tribal god that was at home in ancient Syria, where the Baal of the Heavens, the Lord of Heavens, was worshipped. The sanctuary dedicated to him became a tribal temple. An inscription from Dura Europos identifies Baalshamin with Zeus-Kyrios in the Greek pantheon.[17] Bel and Nabu were Mesopotamian gods, stretching their influence to Syria-Palestine. The Temple of Bel shows traces indicating some fusion of the Mesopotamian god Bel and the local god Baal. The relief of the Palmyrene triad, as in the Temple of Bel, also belonged to the cultic images of the sanctuary dedicated to Baalshamin.

Fig. 8.15 *Granite columns belonging to the portico of the Baths of Diocletian in Palmyra.* Photo: Kenneth Lönnqvist 2004, SYGIS

Fig. 8.16 *The portico and the entrance to the Baths of Diocletian.* Photo: Kenneth Lönnqvist 2004, SYGIS

Fig. 8.17 *The Baths of Diocletian.* Photo: Kenneth Lönnqvist 2004, SYGIS

## 8.3 The Baths of Diocletian

In Roman cities, baths were part of public buildings and places for social gatherings. In Palmyra the baths were situated along the main street, just like the theatre and the *Agora*. The portico of the Baths of Diocletian opens directly to the Grande Colonade (Figs. 8.15, 8.16). As referred to previously, like the *Tetrapylon,* these columns were made of red granite with Corinthian capitals and were erected by Sosianus Hierocles, Governor of Syria, and date from the reign of Emperor Diocletian (AD 284–305).[18]

Only one of the original columns was standing when the others were re-erected in the restoration work in the 1960s. The baths had been mainly built in the 2nd century AD. Corinthian columns surround a pool (Fig. 8.17). The principal pool was entered by stairs; smaller bathing spaces and an octagonal meeting room had been discovered during the Syrian excavations in 1959–1960. There was apparently a *caldarium* for hot baths and *frigidarium* for cooling oneself in the Roman style. Marble was used for covering surfaces in the *opus sectile* technique forming geometrical patterns from slabs of white and other colours.[19]

## 8.4 The Theatre

The theatre of Palmyra is one of the ancient buildings that were originally found partly buried. It is situated on the southern side of the Grand Colonnade near the Temple of Nabu, and its plan follows Roman ideals but Oriental influences are apparent (Figs. 8.18–8.21).

Fig. 8.18 *A panoramic view, the so-called spherical panorama, of the theatre in Palmyra.*
Photo: Gabriele Fangi 2010

Figs. 8.19 *A panoramic view, the so-called spherical panorama, of the theatre.*
Photo: Gabriele Fangi 2010

Fig. 8.20 *Panoramic views, the so-called spherical panorama, of the theatre.* Photo: Gabriele Fangi 2010

The large ruins were first identified in 1895 by Émile Bertone, and some trenches were dug at the site in 1902 with explorations continuing in 1924.[20] Wiegand's expedition documented visible parts of the theatre including the damaged *scaenae frons,* the architectural background of the stage, in 1902 and 1917.[21] The excavations started in 1952. The restoration work took place from 1984 and was executed by Saleh Taha under the direction of Khaled Al-As'ad.[22]

Unlike Greek theatres, the plan of the *orchestra* of Roman theatres is semi-circular. In Palmyra its diameter is 20 metres.[23] The stage is 48 metres wide and 10.5 metres deep.[24] There are 11 sections for 9 existing rows of seats in the *cavea,* and for the significant members of the society, like senators, there were private seats at the front. Scale drawings, spherical photogrammetry, wire-frame models, and virtual reconstructions provide a vision of the space and structure of this ancient theatre (Figs. 8.18–8.25; see Fig. 3.15 and technical details in the appendix).

Edmond Frèzouls has studied the theatre and seen similarities in its location and appearance with the theatre of Petra, another caravan city connected with the nomadic environment. Both theatres are associated with colonnades that separate the urban spaces from the surrounding villages and the nomadic world. In Palmyra there is a special transversal colonnade, also called the Theatre Colonnade (see Fig. 8.22), that surrounds the theatre in a semicircular fashion.[25] The colonnades and theatres appear as integrating elements of the Orient to the Greco-Roman planning unlike Dura Europos that did not have an actual theatre. Frèzouls sees the influence of the Greco-Roman cult of Dionysos at the sites as a defining element for the existence of the theatres.[26] Indeed, Dionysos (equivalent to the Roman god Bacchus), was the god of the theatre, wine, and the underworld. That is well represented in the funerary art of Palmyra as we shall discuss in connection with the funerary structures (Ch. 9).

The *scaenae frons,* the architectural background of the stage, extended 45.27 metres in length (Figs. 8.20, 8.26–8.28),[27] and was imposing and high, with entrances placed in rectangular *exedrae* (recesses), with the *Porta Regia* rising in the centre (Fig. 8.29). ISIS destroyed the structure at the beginning of 2017. The entrances opened onto the stage and were defined by alternating columns and niches in stories. Monolith columns provided a rhythm to the scene. The central entrance had a special and rare decorated baldaquin (ornamental canopy) hanging over, supported by Corinthian columns, and a plafond on its ceiling.[28] There were sculptures in several niches decorating the stage.[29] The rich ornamentation consists of floral patterns, gadroons, and spirals. The structure is reminiscent of the façades of the library of Celsus in Ephesus. Some features call to mind column arrangements found in the tomb façades of Petra. Adnan Bounni and Khaled al-As'ad compare the *scaenae frons* of the Palmyra theatre to the façade of a palace.[30]

As previously mentioned, the excavations of the theatre started in 1952, and the building was restored. Various scholars have suggested slightly different dates for the completion of the theatre, either placing it in the first half or the second half of the 2nd century AD.[31] The connection to the colonnaded street is clear now and apart from the archaeological and stylistic analyses, inscriptions surrounding the area of the theatre have, however, been seen as supporting the dating of the theatre to the 3rd century AD, before Zenobia's revolt.[32] A

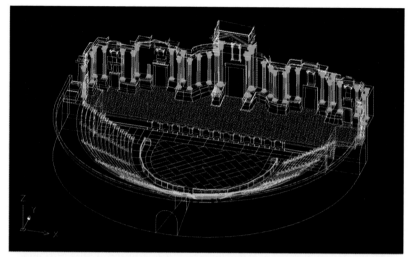

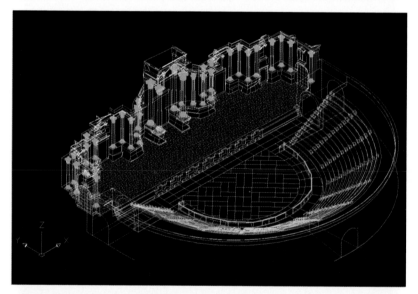

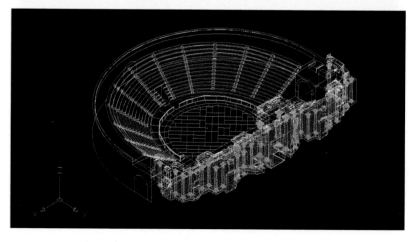

Figs. 8.21–8.25 *Wireframe models of the theatre.* Constructed by Gabriele Fangi and Emanuele Ministri 2016

(Opposite, lower) Fig. 8.26 *Scale drawings of the* scaenae frons. Source: Puchstein, Wiegand 1932

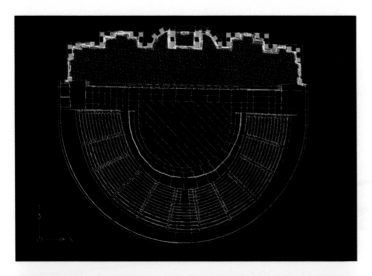

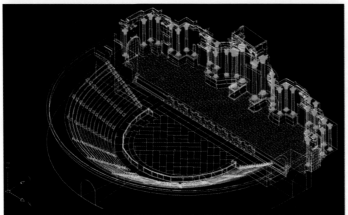

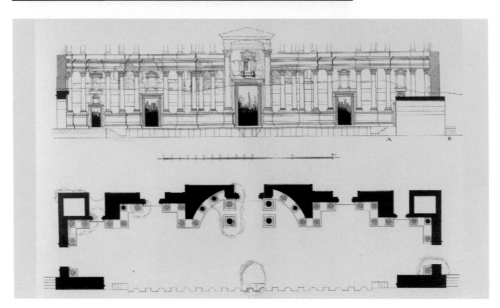

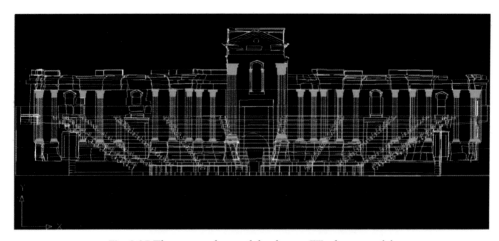

Fig. 8.27 *The* scaenae frons *of the theatre.* Wireframe model
by Gabriele Fangi and Emanuele Ministri 2016

Fig. 8.28 *A 3D model
of the* scaenae frons *of
the Roman theatre in
Palmyra by Gabriele
Fangi and Emanuele
Ministri 2016*

Fig. 8.29 *The Porta Regia
of the theatre.* Photo:
Gabriele Fangi 2010

Fig. 8.30 *A view to the semi-circular transversal colonnade, or the Theatre colonnade, that surrounds the theatre.* Photo: Gabriele Fangi 2010

Fig. 8.31 *A virtual reconstruction of the theatre showing what the building could have looked like in antiquity.* Constructed by Ahmet Denker 2016

Fig. 8.32 *The Agora of Palmyra.* Photo: Kenneth Lönnqvist 2004, SYGIS

Fig. 8.33 *An aerial photograph of the Agora taken in the 1940s.* Source: IFPO

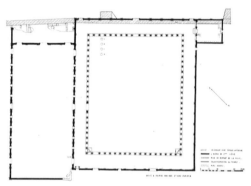

Fig. 8.34 *The plan of the Agora after Seyrig 1940.*

Fig. 8.35 *Earthquake damage visible at the Agora.* Photo: Kenneth Lönnqvist 2004, SYGIS

colonnaded street that adjoins the theatre was erected before AD 254, served as the stage house, as such a structure did not exist in this theatre[33] (Fig. 8.30).

The theatre was the scene of the atrocities of ISIS; the jihadists chose the place for a horrific spectacle. After the recapture of the city from ISIS by the troops of President Assad with the Russians, the Russian Mariinsky Symphony Orchestra performed on its stage in May 2016.[34]

## 8.5 The Agora, the Tariff Court, and the Palmyra Tariff Law

The *Agora* of Palmyra was the commercial centre of the city in the Roman times. Its importance for a caravan city involved in long-distance trade is obvious. Several structures and inscriptions[35] associated with the space reflect its dedication to commercial and administrative interests. The Greek word 'agora' is retained for the site even though the Latin word 'forum' is equivalent. The *Agora* was excavated in 1939 and 1940 by Henri Seyrig, and in 1968 an even larger attached porticoed court 'the annexed space' was excavated. It may have served as a commercial *basilica* (Fig. 8.34 and see map, Fig.3.15).[36]

The rectangular plan of the *Agora* covers 70 metres × 82 metres (Figs. 8.33, 8.34). The site was surrounded by porticoes, the height of which reached nearly 12 metres and had a width of 8 metres running along each of the four sides. The porticoes consist of 80 Corinthian columns. There were refreshing fountains in two corners of the court. They received water from water channels in the city. One fountain in the western corner was associated with a little temple. There were long benches for the banquets that were so common in Palmyra.[37] Dating of the *Agora* has varied. Christiane Deplace and Jacqueline Dentzer-Feydy date it around AD 70, while Michal Gawlikowski follows Jean Starcky's dating to the 2nd century AD.[38]

The Tariff Court (Figs. 8.36, 8.37) and the Senate (Fig. 8.40) are located east of the *Agora*. The original site of the famous Palmyra Tariff or the Tax Law of Palmyra, a large inscription consisting of steles or slabs altogether 5.45 metres in length and 2.80 metres in height, is debatable. This bilingual Tariff was written in Greek and Aramaic and is the largest known Aramaic inscription. It is a decree of taxes levied on merchants and travellers in AD 137. It was transported to Russia during the Ottoman era and the time of the Czars in 1901 as a gift from Ottoman ruler Sultan Abdel Hamid. Now, it is on display in the Hermitage Museum in St. Petersburg (Figs. 8.38, 8.39). The original location of the Tariff has been suggested to be in the Tariff Court or in the vicinity of the 'annexed space,' but recent

Fig. 8.36 *The Tariff Court.*
Drawing by L. F. Cassas

Fig. 8.37 *The Tariff Court.* Photo: Kenneth Lönnqvist 2004, SYGIS

Fig. 8.38 *The Palmyra Tax Law.*
Photo: Kenneth Lönnqvist 2006, SYGIS.
Courtesy of the Hermitage Museum

Fig. 8.39 *One register in the Palmyra Tax Law.*
Photo: Kenneth Lönnqvist 2006, SYGIS.
Courtesy of the Hermitage Museum

excavations place it in the area of the Temple of Rabassire, and a newly found street in the neighbourhood of the *Agora*. Michal Gawlikowski suggests that the street used to be an artery of Palmyra before the Grand Colonnade was built.[39]

Fig. 8.40 *The Senate*. Photo: Kenneth Lönnqvist 2004, SYGIS

# Endnotes

1   Bounni, A., Seigne, J. and Saliby, N. (1992, 2004) *Le Sanctuare de Nabû à Palmyre*. BAH. Tome 131. IFAPO. Paris (1992): Geuthner; Beyrouth (2004): IFAPO; Al-As'ad, Kh. and Yon, J.-B. (2001) *Inscriptions de Palmyre, Promenades épigraphiques dans la ville antique de Palmyre, Guides archéologiques de l'IFAPO*, Beyrouth-Damas-Amman: Institut français d'archéologie du Proche-Orient, p. 50.

2   Drijvers, H.J.W. (1976) *The Religion of Palmyra*, Iconography of Religions XV, 15, Leiden: Brill, p. 19.

3   Wiegand, T., Wulzinger, K. and Schulz, B. (1932) Der korintische Tempel östlich des Theaters, in Wiegand. T., (1932) *Palmyra – Ergebnisse der Expeditionen von 1902 und 1917*, Vol. I, II, Textband, Tafeln, Archeologisches Institut des Deutschen Reiches Abteilung Istanbul, Berlin: Verlag von Heinrich Keller, see Vol. I, pp. 108–121.

4   Bounni, A., Seigne, J. and Saliby, N. (1992, 2004) *Le Sanctuare de Nabû à Palmyre*. BAH. Tome 131. IFAPO. Paris (1992): Geuthner; Beyrouth (2004): IFAPO.

5   Drijvers, H.J.W. (1976) *The Religion of Palmyra*, Iconography of Religions XV, 15, Leiden: Brill, p. 19.

6   Bounni, A. and Al-As'ad, Kh. (1997) *Palmyra: History, Monuments and Museum*, Damascus, pp. 48–53.

7   http://en.unesco.org/news/director-general-irina-bokova-firmly-condemns-destruction-palmyra-s-ancient-temple-baalshamin. Accessed 22nd November, 2016.

8   Wood, R. (1753) *The Ruins of Palmyra, otherwise Tedmor, in the Desart*. London:[Wood].

9   Cassas, L.F. (1798–1804) *Voyage pittoresque de la Syrie, de la Phénicie, de la Palestine et de la Basse Egypte*. Paris: D'Imprimie de la République.

10   Schulz, B. ( 1932) Der Baalsamin-Tempel, in Wiegand. T. (1932) *Palmyra – Ergebnisse der Expeditionen von 1902 und 1917*, Vol. I, II, Textband, Tafeln, Archeologisches Institut des Deutschen Reiches Abteilung Istanbul, Berlin: Verlag von Heinrich Keller, see Vol. I, pp. 122–126.

11   Collart, P. and Vicari, J. (1969) *Le Sanctuare de Baalshamin à Palmyre*, Vol. 1, Topographie et Architecture, Texte, Bibliotheca Helvetica Romana X i, Institut Suisse de Rome, Schwitzerland: Paul Attinger, SA, Neuchâte, p. 5; see also illustrations in Collart, P. and Vicari, J. (1969) *Le Sanctuare de Baalshamin à Palmyre*, Vol. II, Topographie et Architecture, Illustrations, Bibliotheca Helvetica Romana X ii, Institut Suisse de Rome, Schwitzerland: Paul Attinger, SA, Neuchâte.

12   Schulz, B. ( 1932) Der Baalsamin-Tempel, in Wiegand. T. (1932) *Palmyra – Ergebnisse der Expeditionen von 1902 und 1917*, Vol. I, II, Textband, Tafeln. Archeologisches Institut des Deutschen Reiches Abteilung Istanbul, Berlin: Verlag von Heinrich Keller, see Vol. I, p. 123.

13   Browning, I. (1979) *Palmyra*, London: Chatto & Windus, pp. 163–168.

14  Al As'ad, Kh. and Gawlikowski, M. (1997) *The Inscriptions in the Museum of Palmyra*, A *Catalogue*, Palmyra and Warsaw: the Committee for Scientific Research, Warsaw, Poland, pp. 72–73; Collart, P. and Vicari, J. (1969) *Le Sanctuare de Baalshamin à Palmyre*, Vol. II, Topographie et Architecture, Illustrations, Bibliotheca Helvetica Romana X ii, Institut Suisse de Rome, Schwitzerland: Paul Attinger, SA, Neuchâte, Pl. LXI.

15  Collart, P. and Vicari, J. (1969) *Le Sanctuare de Baalshamin à Palmyre*, Vol. 1, Topographie et Architecture, Texte, Bibliotheca Helvetica Romana X i, Institut Suisse de Rome, Schwitzerland: Paul Attinger, SA, Neuchâte, p. 11.

16  Drijvers, H.J.W. (1995) Inscriptions from Allât's sanctuary, in *ARAM*, Vol. 7, pp. 109–119.

17  Seyrig, H. (1937) Comptes rendus, in *Syria*, Vol. 18, pp. 119–120.

18  Browning, I. (1979) *Palmyra*, London: Chatto & Windus, pp. 139–141.

19  Ostrasz, A. (1969) La partie médiane de la rue principale du Palmyre, in *Annales Archéologiques de Syrie*, pp. 109–120; see also Wielgosz, D. (2013) Coepimus et lapide pingere: marble decoration from the the so-called Baths of Diocletian at Palmyra, in *Studia Palmyreńskie*, Vol. XII, pp. 319–332; see further Dodge, H. (1988) Palmyra and the Roman Marble Trade: Evidence from the Baths of Diocletian, in *Levant*, Vol. XX, pp. 215–230.

20  Fourdrin, J.-P. (2009) Le front de scène du théâtre de Palmyre, in *Fronts de scène et lieux de culte dans le théâtre antique*, ed. by Moretti, J.-C., Travaux de la Maison de l'Orient et de la Méditerranée, Vol. 52, pp. 189–233.

21  Puchstein, O. (1932) Das Theater, in Wiegand. T., (1932) *Palmyra – Ergebnisse der Expeditionen von 1902 und 1917*, Vol. I, II, Text, Tafeln, Archeologisches Institut des Deutschen Reiches Abteilung Istanbul, Berlin: Verlag von Heinrich Keller, Text, pp. 41–44.

22  Fourdrin, J.-P. (2009) Le front de scène du théâtre de Palmyre, in *Fronts de scène et lieux de culte dans le théâtre antique*, ed. by Moretti, J.-C., Travaux de la Maison de l'Orient et de la Méditerranée, Vol. 52, pp. 189–233.

23  Browning, I. (1979) *Palmyra*, London: Chatto & Windus, pp. 145–146.

24  Bounni, A. and Al-As'ad, Kh. (1997) *Palmyra: History, Monuments and Museum,* Damascus, p. 74.

25  Frézouls, E. (1961) Récherches sur les théâtres de l'orient syrien, in *Syria*, Vol. 38, pp. 54–86.

26  Frézouls, E. (1961) Récherches sur les théâtres de l'orient syrien, in *Syria*, Vol. 38, pp. 54–86.

27  Fourdrin, J.-P. (2009) Le front de scène du théâtre de Palmyre, in *Fronts de scène et lieux de culte dans le théâtre antique*, ed. by Moretti, J.-C.,Travaux de la Maison de l'Orient et de la Méditerranée, Vol. 52, pp. 189–233.

28  Fourdrin, J.-P. (2009) Le front de scène du théâtre de Palmyre, in *Fronts de scène et lieux de culte dans le théâtre antique*, ed. by Moretti, J.-C.,Travaux de la Maison de l'Orient et de la Méditerranée, Vol. 52, pp. 189–233.

29  Fourdrin, J.-P. (2009) Le front de scène du théâtre de Palmyre, in *Fronts de scène et lieux de culte dans le théâtre antique*, ed. by Moretti, J.-C., Travaux de la Maison de l'Orient et de la Méditerranée, Vol. 52, pp. 189–233; Browning, I. (1979) *Palmyra*, London: Chatto & Windus, p. 145.

30  Bounni, A. and Al-As'ad, Kh. (1997) *Palmyra: History, Monuments and Museum,* Damascus, p. 74.

31  Browning, I. (1979) *Palmyra*, London: Chatto & Windus, p. 145.

32  Fourdrin, J.-P. (2009) Le front de scène du théâtre de Palmyre, in *Fronts de scène et lieux de culte dans le théâtre antique*, ed. by Moretti, J.-C.,Travaux de la Maison de l'Orient et de la Méditerranée, Vol. 52, pp. 189–233, p. 229.

33  Browning, I. (1979) *Palmyra*, London: Chatto & Windus, pp. 145–148.

34  *Reuters* 6th May, 2016.

35  Seyrig, H. (1941) Inscriptions grecques de l'agora de Palmyre, in *Syria*, Vol. 22, pp. 223–270.

36  Seyrig, H. (1940) Rapport sommaire sur les fouilles de l'Agora de Palmyre, in *Comptes rendus des séances de l'Académie des Inscriptions et Belles-Lettres*, Vol. 84, No. 3, pp. 237–249; Deplace, C. and Dentzer-Feydy, J. (2005) *L'Agora de Palmyre*, sur la base des travaux de Seyrig, H., Duru, R. et Frézouls, E., avec la collaboration de Al-As'ad, Kh., Balty, J.-C., Fournet, T., Weber, T.M. et Yon, J.-B., BAH 175. Mémoires. Bordeaux/ Beyrouth, Ausonius: Ifpo.

37  Seyrig, H. (1940) Rapport sommaire sur les fouilles de l'Agora de Palmyre, in *Comptes rendus des séances de l'Académie des Inscriptions et Belles-Lettres*, Vol. 84, No. 3, pp. 237–249.

38  Gawlikowksi, M. (2014) Palmyra: reexcavating the site of the Tariff (fieldwork in 2010 and 2011), in *Polish Archaeology in the Mediterranean,* 23/1 (Research 2011), pp. 415–430.

39  Gawlikowksi, M. (2014) Palmyra: reexcavating the site of the Tariff (fieldwork in 2010 and 2011), in *Polish Archaeology in the Mediterranean,* 23/1 (Research 2011), pp. 415–430.

# 9

# THE TOMBS FOR THE ETERNAL SOULS
# OF PALMYRENES

## 9.1 The Necropoleis – the cities for the dead

In ancient times, the tower tombs guarding the western entrance of Palmyra greeted the caravans and visitors approaching the city. In the Valley of the Tombs, or the so-called western necropolis – a city for the dead – visitors were introduced to and reminded of past members of the ancient families of Palmyra. The most famous towers stood there, but some are also scattered on the plain in the northern necropolis, beneath the Arab citadel, as well in the south-western necropolis.[1] Some of the best-preserved tower tombs, such as that of Elahbel in the Valley of the Tombs, were blown up by ISIS in September 2015. The destruction became visible in satellite images provided by DigitalGlobe.[2]

Fig. 9.1 *The Valley of the Tombs in Palmyra drawn*
*by Louis François Cassas in the 18th century.*

169

Fig. 9.2 *Tower tombs in the Valley of the Tombs in Palmyra photographed by A. Poidebard from the air.* Source: Poidebard, 1934

Fig. 9.3 *The Valley of the Tombs.* Photo: Kenneth Lönnqvist 2004, SYGIS

Fig. 9.4 *Funerary items associated with a ruined tomb including architectural fragments,* loculi, sarcophagi, *and sculpture from Palmyra photographed in 1929.* Photo: Library of Congress, Matson Collection

Tower tombs are the most peculiar and famous funerary structures in Palmyra. They have been depicted in ancient drawings and paintings, such as by Robert Wood[3] and Louis François Cassas[4] (Fig. 9.1), as well as described by travellers such as Charles Greenstreet Addison in 1838 or Gertrude Bell while approaching Palmyra in 1900. Theodor Wiegand's expeditions to Palmyra in 1902 and 1917 also documented tower tombs and other funerary monuments, listing some earlier studies carried out in the *necropoleis* (sing. *necropolis*).[5] Four *necropoleis* were identified by Wiegand's expedition:[6] the Valley of the Tombs or the western necropolis, partly belonging to the south-western necropolis, the northern or north-western necropolis, the southern or the south-eastern necropolis, the latter can also refer to two separate *necropoleis* (see the plans in Ch. 3, Figs. 3.15, 3.18, 3.19).[7]

The tower tombs are important monuments that characterise the ancient city and express its cultural identity. They form a striking feature in the built landscape, typical for this oasis city, demonstrating the beliefs of its people in the afterlife, and the importance of Palmyrene families, tribes, and clans. They commemorate the importance of Palmyra's society. Nowhere else in the Greco-Roman world are such tower-like funerary monuments seen in quite the concentration and numbers of Palmyra. At the ground level, there are other types of funerary buildings too: *mausolea* (sing. *mausoleum),* some representing house- or palace-like tombs, and funerary temples.[8] As well as the ground-level monuments, there are underground tombs, namely *hypogea* (sing. *hypogeum*), and grottos cut into cliff faces.[9] The *hypogea* are found in the Valley of the Tombs,[10] and the south-western[11] and southern/south-eastern *necropoleis.*[12] In the region of the Camp of Diocletian there also are tombs that are more like grottos.[13] In some cases, as in Petra, ancient tombs have been reused as habitation areas for the Bedouins. Clearly, different types of elaborate tombs belong to the prime of Palmyra dating from the 1st century BC to the 3rd century AD. Altogether some 300 funerary monuments have been identified from Palmyra, of which 180 are tower tombs.[14]

*Fig. 9.5 Overturned sarcophagi photographed in Palmyra in 1929.* Library of Congress, Matson Collection

Fig. 9.6 *A sarcophagus that stood in the entrance of the Palmyra Museum.* Photo: Silvana Fangi 2010

Fig. 9.7 *Looted tombs on Jebel Bishri in the Palmyrides, not far from Halabiya.*
Photo: Minna Lönnqvist 2004, SYGIS

Besides the monuments and underground chambers for the wealthy of Palmyra there were naturally more simple graves. In addition to the previously mentioned cairns/*tumuli* found around Palmyra and its nomadic territory (Ch. 2 and 4), there were other modest funerary structures that consisted of steles erected on earthen graves during the Greco-Roman period. Many such steles were preserved and displayed in the rooms and gardens of the Palmyra Museum.[15] Long cairns and steles are also common markers on Muslim graves in the area today, often located on ancient tells (see, for example, Ch. 3, Fig. 3.1).

*Loculi* (sing. *loculus*) – wall niches for *sarcophagi* or ossuaries – appear in tower tombs, *hypogea*, and grottos. A *sarcophagus* is a coffin made of hard material like stone (Figs. 9.4–9.6), terracotta, or metal. Ossuaries, also made of hard material and often decorated in relief, are smaller chests for bones used in secondary burials. In Palmyra, *loculi* are closed with stone slabs on which portraits of the deceased are cut in relief. On the walls of the tombs there also may appear *arcosolia* (sing. *arcosolium*), semi-circular niches for bodies, *sarcophagi*, or large slabs decorated in relief. As we have already seen, funerary art in Palmyra was especially lavish. It shows how the dead were valued in the society, and the belief in the triumph of death and life continuing after the death.

As far as the ancient tombs are concerned, Palmyra and its surrounding areas have been under continuous and heavy looting. Louis François Cassas already illustrated the world of tomb raiders in Palmyra in the 18th century. Gertrude Bell's photographs also depict robbers' diggings in the plain of Palmyra in 1900.[16] Looted tombs and even tomb robbers were encountered by the present principal author in 2005 in the region of Halabiya while studying the district (Fig. 9.7).

## 9.2 Towers as funerary structures

Polygonal towers became popular architectural structures in the Hellenistic Age. The Lighthouse of Pharos in Alexandria became a tower *par excellence* and was among the Seven Wonders of the Ancient World.[17] Interestingly, the areas that Alexander the Great conquered included staircase towers in temples and tower tombs stretching as far away as Arabia and India.[18] Tower tombs of various kinds appear throughout Persia and the Hellenistic and Roman Near East, and in the ancient world in general. Even in South America where they are known as *chullpas*.[19] However, the architectural shapes and dates vary. The architectural influence of Persia and its temples, including fire altars, have been discussed as possible architectural sources of inspiration for the tower tombs, like those of Palmyra.[20] The ideas and the influence of the form seem to be consistent with the strong Parthian impact in the region and the connections to the Silk Road.

Fig. 9.8 *Tower tombs in the Valley of the Tombs in Palmyra photographed in 1920–1933.* Photo: Library of Congress, Matson Collection

Comparable tower tombs to those in Palmyra exist around the Euphrates valley in the areas that were under the influence of Palmyra: up river in Edessa (modern Urfa), Qalaat Jaber near Sura on the Euphrates, on the peninsula of Halabiya, at Tabus, at Dura Europos, at Baghouz, and at Al-Qaim near the Iraqi border. Hauran in Syria also has some tower tombs that are

Fig. 9.9 A 3D model of rock-cut
tombs from the Greco-Roman
period in Petra in Jordan. Façades
of tower-like tombs. Courtesy:
The Zamani Project

Fig. 9.10 The Jerusalem necropolis in the Kidron Valley of Jerusalem in Israel. On the left behind a tree
stands the so-called Absalom's tomb, on the right rock-cut tomb of the Benei Hezir and a monolith
pyramidical tomb of Zacharias, all from the Greco-Roman period. Photo: Minna Lönnqvist 2012

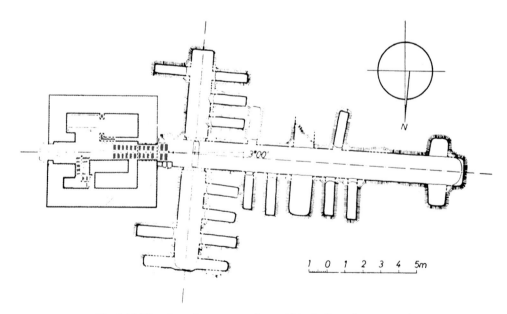

Fig. 9.11 The plan of a tower tomb no. 15 including a hypogeum in
the Valley of the Tombs in Palmyra. After K. Michalowski 1964

similar to those in Palmyra, as well as Cilicia in Turkey.[21] However, there are some technical and decorative differences among these structures on the Middle Euphrates, Hauran, and Cilicia compared to those in Palmyra itself.

Gertrude Bell also recorded and photographed two tower tombs in Serrin near Tell Ahmar in Mesopotamia. Local Arabs called them 'windmills'. An Aramaic inscription dates one to AD 74. It was built by a man called Manu for himself and his sons. Bell surmised that these towers were topped by pyramids, unlike the majority of those in Palmyra and Hauran. In addition, Bell identified another tower tomb and photographed it at Neshabah in Ka'lat Jab'ar. There the tombs appear to have been located on hills,[22] offering visibility from them and to them from afar. These tower tombs had pilasters on the exterior wall like the ones at Dura Europos and Tabus.[23]

Beside Palmyra, Hatra in Iraq has tower tombs inserted into the walls of the city.[24] In Persia there are similar structures that can be seen in the rock-cut façades of Petra in Jordan (Fig. 9.9). These are all famous caravan cities that flourished at the same time and had lively contacts. In Jerusalem, the Kidron Valley has some comparable features in its tombs. Absalom's tomb and Zachariah's tomb are tower tombs, the former capped with a pointed peak like the rock-cut tombs in Petra and the latter by a pyramid.[25] It is possible that some tombs in the Jerusalem necropolis originally had towers in their façades but the structures have eroded away, as Dan Barag has suggested for example, in the case of the tomb for Benei Hezir in the Jerusalem necropolis (Fig. 9.10).

In the Aramaic culture, funerary monuments appearing above ground, such as towers, were called *nafsha*, referring to a commemorative monument for a soul, and still the word *nefesh* is used in Hebrew for such monuments (see, for example, Fig. 9.34). Herod the Great also built towers in the Citadel of Jerusalem, naming them after his relatives.[26] Some tower tombs of Palmyra, including tomb no. 15 in the Valley of the Tombs, also incorporate *hypogea*, thus providing underground chambers (see Fig. 9.11).[27]

(Left) Fig. 9.12 *Early tower tombs in the Valley of the Tombs photographed in 1920.* Photos: Library of Congress, Matson Collection

(Right) Fig. 9.13 *An early tower tomb in the Valley of the Tombs photographed in 1920.* Photos: Library of Congress, Matson Collection

Inhumation was part of the burial customs, including mummification, which took place in Palmyra. Inhumation came from the Semitic funerary tradition differing, from the ancient Indo-European Greek and Roman custom of cremation. There were also secondary burials that meant collecting the bones of the deceased after the decomposition of a body and burying them. Mummification seems to have been brought to Palmyra through the contacts with Egypt. Semitic and Hamitic cultures wished to preserve the deceased, at least their bones, to mark their importance. Some scholars, for example Ernst Herzfeld and Karl Watzinger, have suggested that the tower tombs even imitate inhabited towers, making them houses for the dead. They were seen as a way of building structures in the way of Arab nomads within the limits of urban areas.[28] Tombs with names such as 'House Alfa' could have served as eternal houses. In addition, the Temple of Bel in Palmyra and Temple of Jupiter in Baalbek in Lebanon incorporate staircase towers for astral cults;[29] and the tower tombs could have served such beliefs as well.

Fig. 9.14 *The drawing of a tower tomb from Palmyra published by Wood in 1753.*

The identification of the families and the deceased in the tombs has been based on the inscriptions found, usually above the entrance and in the graves (see Fig. 9.24). There have been studies that have traced the evolution of tower tombs on typological grounds based on variations in shapes as well as decorations. Also the types of portraits of the deceased can be used for relative dating. The types of the tower tombs have further been reorganised in a relative chronological order by Michal Gawlikowski and Agnes Henning. The acme of the tower tombs appears to have been in the 1st century AD in particular.[30] Unlike the typological relative dating, inscriptions can provide exact foundation dates for some of the structures. Also, coins and stamps, if found in the graves, may provide precise *terminus*-dates.

The earliest tower tombs have a very basic structure either comprising a distinct larger *podium* or without any section for a base or any elaborate decorations (see Figs. 9.12, 9.13). The size of the ground-plan is generally modest, sometimes covering 7 metres × 5 metres. The stone blocks used for construction are irregular. The *loculi*, special rectangular niches in the walls, were in the first phase set out on the exterior walls and inside the tombs. Ernest Will has also divided such *loculi* into various types that may have straight walls and ceilings, like in Palmyra, and blocked in at the end with a funerary slab. (In Dura Europos there were also pyramidal pointed ceilings for *loculi*. Comparable *loculi* were found in Tabus on Jebel Bishri by the project SYGIS led by the present principal author in the Palmyrides.)[31] There are no staircases for various floors in this first type. These tombs form a compact group of their own in the Valley of the Tombs. The date for this first group has been determined as 1st century BC, being partly Hellenistic.[32]

The second type of the tower tombs in Will's typology is larger, elevated with a different stepped basement, and incorporates several floors, with the entrance facing the valley and a staircase that connects the floors. The ashlars are still irregular. The *loculi* are inside the building

Figs. 9.15, 9.16 *Elahbel's tower tomb in the Valley of the Tombs photographed in 1900–1920 before its restoration.* Photos: Library of Congress, Matson Collection

arranged on the walls in central burial chambers. Three tower tombs belong to this group: those of Atenatan, Hairan, and Kithot. All date from the first half of the 1st century AD. A good example is the tower tomb of Kithot that dates to AD 44.[33] In this period the windows and/or balconies start appearing on outer walls. Gawlikowski has studied tombs belonging to this group. The typology of the plans and the location of staircases vary in this and the following group.[34]

Agnes Henning sees that a notable change in the typology and furnishing of the tower tombs was taking place in the second half of the 1st century AD and associated with the urban development of Palmyra.[35] In Will's typology this is the time of the third group that consists of regular forms and the most elaborately decorated tombs such as the tower tombs of Jamblique/Iamblichus and Elahbel, the first dating to AD 83 and the second to AD 103. (See the Tomb of Elahbel in Figs. 9.15, 9.16, 9.26-9.34). Both have elevated and strong bases that are generally constructed from larger stones than those used in the upper parts. In this type there is sometimes a balcony or a window on exterior walls (Fig. 9.14).[36] Those could have been filled with sculptures. The tombs of Jamblique/Iamblichus and Elahbel have frequently been documented in drawings and in early photographs (Figs. 9.14, 9.16) including their elaborate interiors. Apart from reliefs and funerary sculpture there are generally murals decorating interior walls of some tombs. The motifs consist of ribbons, floral designs, and decorations.[37]

The plan of Elahbel's tomb covers 12.35 metres × 12.35 metres, so it is larger than the early tower tombs.[38] These later tower tombs have fine interiors with several floors connected by a staircase, even entrances to a roof terrace. Plaster decorations like *stucco* in ceilings and walls may be rich, and the walls may have Corinthian pilasters. Robert Wood published decorations

Figs. 9.17–9.24 *Modelling Elahbel's Tower Tomb in 3D.*
Models by Gabriele Fangi and Marco Franca, 2016

from the interiors of Elahbel's tomb in 1753.[39] Beautiful colours were used, like blue in the ceiling of the tomb[40] (Figs. 9.26–9.32). The Tower Tomb of Elahbel had been photogrammetrically measured by A. Brall and his students[41] and digitally documented and remodelled in 3D by Gabriele Fangi and his students (see Figs. 9.17–9.24) before ISIS blew it up in 2015. Also Rekrei has modelled both the tomb of Elahbel and the tomb of Iamblichus.[42]

Beside the funerary function Will hinted at another function of the tower tombs that we have also suggested because of the structure of some tombs and their location in the landscape along the roads and entrance sections of cities such as Palmyra, Dura Europos, and Halabiya. The observational and defensive use of these tombs can also be suggested because they have been inserted in defensive walls in Palmyra and Hatra. As previously mentioned, in Palmyra in some cases they have upper platforms, balconies, and windows to carry out observations. The placement of structures in the terrain thus utilising the topography, and providing views of afar speak for their multifunctional purposes.[43]

The tombs were houses for eternity. However, as the world has vividly seen in the destruction by ISIS, the monuments are not eternal.

Figs. 9.26, 9.27 Interiors of Elahbel's tower tomb with Corinthian pilasters photographed approximately 1920–1933. Library of Congress, Matson Collection

Fig. 9.28–9.31 *Interior decorations from Elahbel's tomb*. Photos:
Kenneth Lönnqvist 2004, SYGIS, and Gabriele Fangi 2010

(Left) Fig. 9.32 *Cassette ceiling with blue colours in Elahbel's tomb.* Photo: Silvana Fangi 2010

(Below) Fig. 9.33 *A funerary sculpture from Elahbel's tomb.* Photo: Bernard Gagnon, 2010

(Above) Fig. 9.34 *The funerary sculpture of a male from Elahbel's tomb lying on a couch.* Photo: Kenneth Lönnqvist 2004, SYGIS

(Right) Fig. 9.35 *The so-called Absalom's tomb in the Jerusalem necropolis dating from the Greco-Roman period. The tower-like structure is called* nefesh *and has common features in pilasters with the tower tombs in the Euphrates valley. The hole in the wall of the monument indicates to robbers' work.* Photo: Minna Lönnqvist 2012

## 9.3 A palace, house tombs, and funerary temples

As previously mentioned, there are also imposing palace-like structures, houses, and funerary temples, real *mausolea,* in Palmyra. The purpose seems to have been to imitate real houses, such as *peristyle* houses (see Ch. 4 and Fig. 9.38), which are common types in Palmyra. One palace-like structure is in the centre of the Valley of the Tombs, identified as tomb no. 36.[44] Its plan covers 18 metres × 18 metres. The tomb dates to AD 210–230. Altogether 300 graves were included in the *loculi* of this tomb, and it has been suggested that the tomb belonged to the family of Iulius Septimius Aurelius Vorodes, a famous dignitary in Palmyra. The lavish ornamentation on the grave structure has been shown to represent local craftsmanship that might have its inspiration in local textile work. Dionysos-Baalshamin, Nereids, and Erotes are illustrated;[45] the themes with the vine are apparently referring here, in the funerary context, to Greek traditions associated with the underworld and rebirth, that is ideas of the afterlife.

The House of Marona (Fig. 9.36), the tomb of a Palmyrene trading family that was involved in sea trade, is situated north of the city wall. In a rectangular house there are Corinthian corner pilasters appearing to support the ceiling. In the south-eastern pilaster there is a snake as a carved decoration and that is why the Arabic name of the tomb is Qasr el Hayye, Palace or Castle of the Serpent. The tomb had been originally built by Julius Aurelius Marona in AD 236, reused as a house, and because of its partial collapse into ruins it was re-erected in 1946.[46]

Several funerary temples have been identified in the northern necropolis.[47] The most famous example is found near the Grand Colonnade, the location therefore providing visibility and underlining the spatial importance of the building. Its façade differs from the *mausolea* mentioned above. There is a portico in front like those in Greco-Roman temples, and therefore it is closer to a temple than the palaces mentioned above. This funerary temple, identified as tomb no. 86, also dates from the 3rd century AD. Earlier just six Corinthian columns of the portico in the front or columns *in antis* were standing (Fig. 9.37), but walls were later raised during restoration work. The plan covers 18.5 metres × 15.5 metres. Three burial spaces filled with *loculi* surround the walls and in the centre there are four columns, between which there could have been a *sarcophagus.* In the pilasters we see again lively vine carvings. In the back of the space there is an entrance to an underground

Fig. 9.36 *The House of Marona tomb photographed in Palmyra in 1920.* Library of Congress, Matson Collection

Fig. 9.37 *The portico of the Funerary Temple in Palmyra photographed in 1867.* Library of Congress, Maison Bonfils

Fig. 9.38 *The plan for the peristyle type of house tomb in the south-eastern necropolis.* Source: Gabriel 1926

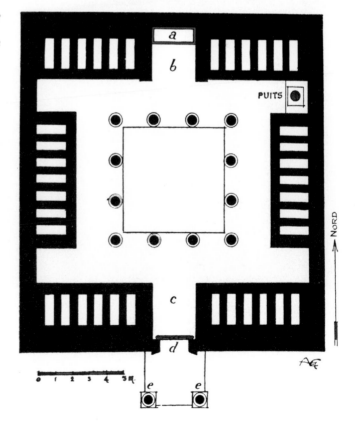

Fig. 9.40 *The entrance to the Tomb of Dionysos in the south-western necropolis photographed sometime in 1920–1933. Note the door visible as opened.* Photo: Library of Congress, American Colony collection

Fig. 9.39 *The entrance with an original limestone door to a* hypogeum *in the south-eastern necropolis.* Photo: Minna Lönnqvist 2000

Fig. 9.41 *3D model of the* hypogeum *tomb H in the south-eastern necropolis.* Courtesy: K. Saito

crypt that also contains *loculi.*[48] Corinthian pilasters appeared in the interior as in the tower tomb of Elahbel, between which there were portraits of the deceased at the end of the *loculi.* Wiegand has made an impressive hypothetical reconstruction for this funerary temple.[49]

## 9.4 Hypogea – underground tombs

The most famous *hypogea* of Palmyra are those of Yarhai (Figs. 9.42, 9.43), situated in the Valley of the Tombs,[50] and the Tomb of the Three Brothers (Figs. 9.49, 9.50) in the south-western necropolis.[51] The first one was later set up into an exhibition at the National Museum in Damascus (Fig. 9.44). Similar *hypogea* have been excavated in recent decades by Japanese teams in the south-eastern part of the city. The majority of the *hypogea* have been decorated in relief but some also have beautiful frescos painted on the walls and ceilings.

In the underground tombs there may be enormous limestone doors, like those in palaces, through which visitors can enter with large keys. Kiyohide Saito and a Syro-Japanese team have entered and studied *hypogea* with such doors. The present principal author visited some of those *hypogea* with her research team in 2000. The large stone doors are decorated by carved coffered panels (see Fig. 9.39). Inside the dark tombs one finds the world in which the deceased are still continuing their life by taking part in funerary banquets. They lie on the *kline* type of sofa on stone *sarcophagi* surrounded by their family members and wearing fine embroidered clothes. The Ancient Egyptian Song of Harpists called on listeners to be cheerful and feast on the day of a funeral as everybody is mortal.[52] This attitude is also found in Greco-Roman epitaphs.

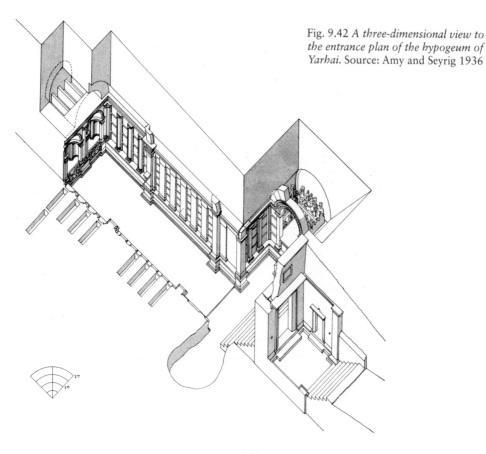

*Fig. 9.42 A three-dimensional view to the entrance plan of the hypogeum of Yarhai. Source: Amy and Seyrig 1936*

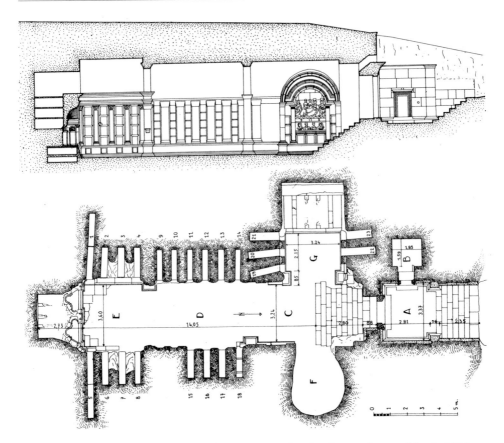

Fig. 9.43 *Long section and plan of the hypogeum of Yarhai.* Source: Amy and Seyrig 1936

Fig. 9.44 *The western exedra with graves closed and marked with portraits, with a banqueting scene placed in the middle, from the hypogeum of Yarhai now reconstructed and exhibited at the Damascus Museum.* Photo: Kenneth Lönnqvist 2004, SYGIS

Fig. 9.45 *The western exedra from the hypogeum of Yarhai as a scale drawing.* Source: Amy and Seyrig 1936

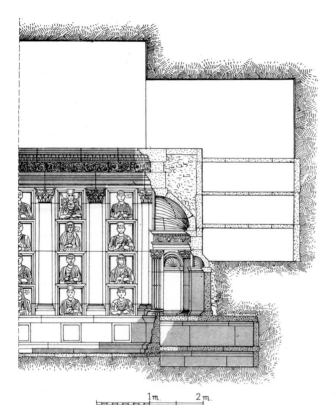

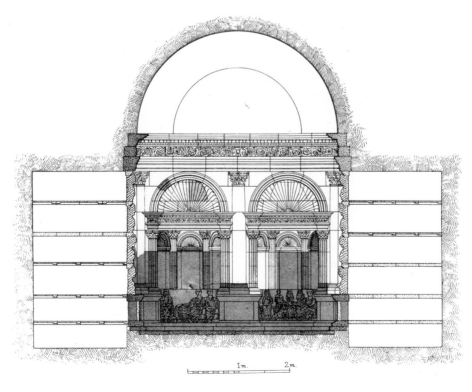

(Left) Fig. 9.46 *Funerary sculpture in the* hypogeum *of Yarhai.* Source: Amy and Seyrig 1936

(Below) Fig. 9.47 *The southern* exedra *of the* hypogeum *of Yarhai.* Source: Amy and Seyrig 1936

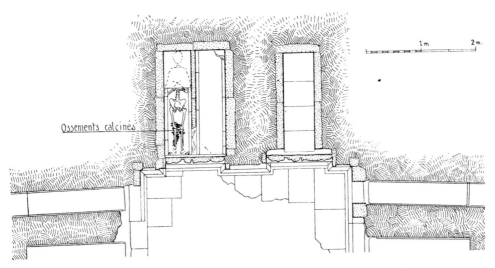

Ossements calcinés

1 m    2 m.

Fig. 9.48 *A deceased layed in a loculus in the tomb of Yarhai.* Source: Amy and Seyrig 1936

Fig. 9.49 *The* hypogeum *tomb of the Three Brothers;* sarcophagi *and* loculi, *a tomb chamber photographed in 1920–1933.* Photo: Library of Congress, the American Colony collection

The building of the *hypogeum* of Yarhai (Figs. 9.42–9.48) was a considerable undertaking. The tomb dates from AD 108–240. The tomb had been visited by robbers before the excavations of the 1930s. The entrance of the door faces north, and the tomb is accessed by steps that lead to the entrance hall. Here we have a limestone door of the type that has been described above, but this door is small. The length of the tomb is over 20 metres and the width is around 15 metres at the widest point. There is a main gallery divided into three spaces flanked by three walls with *exedrae* (semicircular recesses). The tomb was hewn in limestone but has an earthen floor covering its main areas. The chief gallery is simpler than the *exedrae*, which are richly decorated. Corinthian pilasters, rich floral architectural decorations, and sculptures of the deceased are of highest quality. Here again several scenes of the deceased banqueting were found. The drinking cups of the *skyphos* type are richly decorated, for example, with vine ornamentation. Dionysiac/Bacchic figures appear in the architectural decorations. There were bilingual inscriptions in Palmyrene Aramaic and Greek. Approximately 219 people were buried in the tomb, along with *unguentaria*, bottles for ointments, reflecting the funerary custom in which bodies were treated with expensive ointments. Lamps were included in the funerary rituals as well and also had the practical use of seeing in the darkness of the tomb.[53]

The Tomb of the Three Brothers (Figs. 9.49, 9.50), situated in the southern entrance section of the city, is exceptional with its beautiful and colourful frescoes. The inscriptions in the entrance of the tomb tell us that it belonged to three brothers Naʾamai, Malay, and Saʾalai, and the use of the tomb is dated to AD 160, 191, and 241. Beautiful frescoes with geometrical motifs and medallions with portraits inside are supported by winged Nikes or Victories – goddesses from the Greco-Roman world. Other Greek mythological scenes, among them motifs from the Trojan War such as Achilles, are included.[54] The Tomb of the Three Brothers with its valuable frescoes had been recently under the care of the World Monument Fund.[55]

Fig. 9.50 *Frescoes from the central burial chamber including the Rape of Ganymede and portraits in medallions in the Tomb of the Three Brothers photographed in 1920–1933.* Photo: Library of Congress, the American Colony collection

## 9.5 The grotto of the Family Alainê

The large tomb of the family Alainê is situated in the cliff face near the Camp of Diocletian. It underwent various phases of construction, starting in AD 138. The second century was a time of great prosperity for Palmyra. Besides *sarcophagi*, portraits, and other sculptures, several small finds explain the character of the owners and the use of the tomb.[56]

Here we see also, beside Palmyrene gods Malakbel and Aglibol, the impact of the Greek god Dionysos in architectural decorations as well as in the embroidery of clothes. The Greek god Eros and *amorini* appear as well.[57] Here the fusion of local Near Eastern and Greek elements of beliefs is visible, as commonly seen in the Orient from the Hellenistic times onwards. The vine is a very common subject in ornamentation in tombs in the Greco-Roman world, and from the Greek world this is associated with Dionysos, god of the underworld, vegetation, and theatres. Garlands with *putti* or *amorini* inserted among the floral scrolls are also depicted. The appearance of the Eros and *amorini* refer to the rejuvenation of life in the underworld.

A group of *penteklinium*, namely five funerary beds or couches like *klinai,* were found in this family tomb. As in several tombs the deceased are represented feasting and wearing luxurious clothes. The detail of the embroidery and the shoes are magnificently presented.[58]

## 9.6 Mummies and mummification in Palmyra

It is most interesting that the funerary customs of the Palmyrenes included artificial mummification techniques of the deceased that closely resemble the customs of the Egyptians. This elaborate conservation of deceased for the afterlife in Palmyra was reserved for rich persons. The mummification in Palmyra was associated with the building of the tower tombs from the 1st century BC to the 3rd century AD.[59] As seen, connections with Egypt were dynamic during the life of Alexander the Great and his successors the Seleucids of Syria and the Ptolemies of Egypt in the Hellenistic period. Interestingly, both Egypt and Syria were concentrating on solar and astral worship in the Greco-Roman times.

In Palmyra, the first known discoveries of mummies were made by the Umayyads in the 8th century AD.[60] They were Arab rulers who, after the Islamic conquest, built castles such as Qasr al-Hayr al-Gharbi and Qasr al-Hayr ash-Sharqi at older Roman sites in the desert in the vicinity of Palmyra (see Ch. 10).[61] The first mummy that they discovered was a female who wore no fewer than 60 pieces of jewellery of various kinds. One hand of the mummy was detached because of the weight of the jewellery. This was a rare find as the deceased of Palmyra were not usually adorned with jewellery, although the portraits depict them in lavish amounts. Dozens of mummies were later found, especially in the excavations led by Robert Amy and Henri Seyrig in the 1930s and 1940s in the tower tombs in the Valley of the Tombs. Mummies have been found in excavations, mostly in the tower tombs of Elahbel and Jamblique/Iamblichus. As late as 1993 a mummy was found in the tower tomb of Atenatan.[62]

The Egyptian mummification process has been vividly described by ancient Greek historian Herodotos.[63] There was the process of removing the organs, washing the corpse, using various chemical substances like asphalt, salts, incense, and perfumes for conservation, and finally rolling the corpse in bandages, usually made of linen. In Palmyra there were two methods for the removal of organs: either following the Egyptian style, taking the brain out through a hole in the skull, or alternatively, totally removing the head of the mummy. After extraction of the brain, other viscera were removed with an instrument. The Palmyrenes left genitals intact. Unlike the Egyptians the Palmyrenes did not preserve the organs in canopic vessels, but rather it seems the organs were burned. The corpses were covered in natron (hyrdrous sodium carbonate) for 70 days in Palmyra, as in Egypt, but the methods of drying varied. Bandages were added.[64]

In the 19th and 20th century, many mummies from Egypt, and in the 20th century from Palmyra, were taken or sold to western countries. Some ended up in museums and some were taken to private collections. A famous example of the Palmyra mummies is in the collection of Carlsberg's Glyptothek in Copenhagen in Denmark. After 1945 the mummies of Palmyra largely faced oblivion. The National Museum of Damascus has had around ten examples, and a few remained in the Museum of Palmyra. The conservation of these mummies in Damascus and Palmyra has been a practical problem, and insects and fungi have taken over in various cases.[65]

An exhibition room in the Palmyra Museum was dedicated to three conserved mummies in 2005. All the mummies were destroyed in a most desecrating way by ISIS in 2015–2016, who took them out from the museum and drove over them.

## Endnotes

1 Watzinger, C. and Wulzinger, K. (1932) Die Nekropolen, in Wiegand, T. (1932) *Palmyra – Ergebnisse der Expeditionen von 1902 und 1917*, Vol. 1, Text, Archäologisches Institut des deutschen Reiches, Abteilung Istanbul, Berlin: Verlag von Heinrich Keller, pp. 45–76.

2 BBC, 4 September, 2015; the *necropoleis* are discussed by C. Watzinger and K. Wulzinger, in Wiegand, T.(1932) *Palmyra – Ergebnisse der Expeditionen von 1902 und 1917*, Vol. 1, Text, Archeologisches Institut des deutschen Reiches, Abteilung Istanbul, Berlin: Verlag von Heinrich Keller, pp. 45–76, and the tower tombs are dealt with by C. Watzinger in pp. 77–84.

3 Wood, R. (1753) *The Ruins of Palmyra, Otherwise Tedmor in the Desart*. London: [Wood]

4 Cassas, L.F. (1789–1804) *Voyage pittoresque de la Syrie, de la Phénicie, de la Palestine et de la Basse Egypte*. Paris: D'Imprimie de la République.

5 Wiegand, T. (1932) *Palmyra: Ergebnisse der Expeditionen von 1902 und 1917*, Vol. 1, Text, Archäologisches Institut des Deutschen Reiches, Abteilung Istanbul, Berlin: Verlag Heinrich Keller, see the list of earlier sources on p. 46.

6 Watzinger, C. and Wulzinger, K. (1932) Die Nekropolen, in Wiegand, T. (1932) *Palmyra –Ergebnisse der Expeditionen von 1902 und 1917*, Vol. 1, Text, Archäologisches Institut des deutschen Reiches, Abteilung Istanbul, Berlin: Verlag von Heinrich Keller, pp. 45–76.

7 Watzinger, C. and Wulzinger, K. (1932) Die Nekropolen, in Wiegand, T. (1932) *Palmyra –Ergebnisse der Expeditionen von 1902 und 1917*, Vol. 1, Text, Archäologisches Institut des deutschen Reiches, Abteilung Istanbul, Berlin: Verlag von Heinrich Keller, pp. 45–76; Henning, A. (2010) The tower tombs of Palmyra: chronology, architecture and decoration, in *Studia Palmyreńskie*, Vol. 12, pp. 159–176.

8 See a general introduction, for example, in Browning, I. (1979) *Palmyra*, London: Chatto & Windus, pp. 172–179.

9 Yon, J.-B. (2002) *Les notables de Palmyre, Études d'histoire sociale*, (BAH) 163, Beyrouth: IFAPO, p. 197; Anadol, S. (2008) Palmyra – Identity Expressed through Architecture and Art, in *Jebel Bishri in Context: Introduction to the Archaeological Studies and the Neighbourhood of Jebel Bishri in Central Syria*, ed. by Lönnqvist, M., BAR International Series 1817, Oxford: Archaeopress, pp. 59–72; Sadurska, A. (1977) *Palmyre VII, Le tombeau de famille de 'Alainê*, Varsovie: Éditions scientifiques de Pologne, p. 17.

10 See, for example, Browning, I. (1979) *Palmyra*, London: Chatto & Windus, pp. 198–201.

11 See, for example, Bounni, A. and Al-As'ad, Kh. (1997) *Palmyra: History, Monuments and Museum*, Damascus, pp. 98–100.

12 Higuchi, T. (1994) *Tombs A and C Southeast Necropolis Palmyra Syria Surveyed in 1990–1992*. Publication of Research Center of Silk Roadology, Vol. 1, Nara, Japan: Research Center of Silk Roadology; Saito, K. (2004–2005) Excavations at the Southeast Necropolis in Palmyra, in *Les Annales Archéologiques Arabes Syriennes*, Vols. XLVII–XLVIII, pp. 137–149.

13 Sadurska, A. (1977) *Palmyre VII, Le tombeau de famille de 'Alainê*, Varsovie: Éditions scientifiques de Pologne, pp. 17–19.

14 Henning, A. (2010) The tower tombs of Palmyra: chronology, architecture and decoration, in *Studia Palmyreńskie*, Vol. 12, pp. 159–176.

15 Yon, J.-B. (2002) *Les notables de Palmyre, Études d'histoire sociale*, BAH. Tome 163, Beyrouth: IFAPO, p. 197; see also Al As'ad , Kh. and Gawlikowski, M. (1997) *The Inscriptions in the Museum of Palmyra, A Catalogue*. Palmyra and Warsaw: The Committee for Scientific Research, Warsaw, Poland.

16 See Silver, M., Rinaudo, F., Morezzi, E., Quenda, F. and Moretti, M.L. (2016) The CIPA Database for Saving the Heritage of Syria, in *ISPRS Archives*, Vol. XLI-B5, pp. 953–960, especially p. 959, Fig. 19.

17 See Fedak, J. (1990) *Monumental Tombs of the Hellenistic Age: A Study of Selected Tombs from the Pre-Classical to the Early Imperial Era*. Toronto: the University of Toronto Press.

18 See Amy, R. (1950) Temples à Escaliers, in *Syria*, Vol. XXVII, pp. 82–136; see also Potts, D. (1990) *The Arabian Gulf in Antiquity*, Vol. II, From Alexander the Great to the Coming of Islam, Oxford: Clarendon Press, pp. 264–266.

19 See, for example, Kesseli, R., Liuha, P., Rossi, M. and Bustamante, J. (1999) Archaeological and Geographical Research of Precolumbian (AD 1200–1532) Grave Towers of Chullpa on the Bolivian High Plateau in the years 1989–1998, Preliminary report, in *Dig it all, Papers Dedicated to Ari Siiriäinen*, ed. by Huurre, M., Carpelan, C., Halinen, P., Kirkinen, T., Laulumaa, V., Lavento, M. and Lönnqvist, M., Helsinki: Archaeological Society of Finland, Finnish Antiquarian Society, pp. 335–348.

20 See the discussion in Potts, D.T. (2010) *Mesopotamia, Iran and Arabia from the Seleucids to the Sasanians*, Farnham, Surrey, England: Ashgate Publishing Limited, pp. 271–300.

21 Will, E. (1949a)  La tour funéraire de Palmyre, in *Syria*, Vol. 26, pp. 87–116; Silver, M., Törmä, M., Silver, K., Okkonen, J. and Nuñez, M. (2015) The Possible Use of Ancient Tower Tombs as Watchtowers in Syro-Mesopotamia, in *ISPRS Annals* (II-5/W3), ed. by Yen, Y-.N.,Weng, K.-H. and Cheng, H.-M. pp. 287–293; Will, E. (1949b) La Tour Funeraire de la Syrie et les Monuments Apparentés, in *Syria*, Vol. 26, p. 311.

22 Bell, G. (1911) *Amurath to Amurath*, London: William Heineman,  p. 30, pp. 36–38, p. 46.

23 Silver, M., Törmä , M., Silver, K., Okkonen, J. and Nuñez, M. (2015) The Possible Use of Ancient Tower Tombs as Watchtowers in Syro-Mesopotamia, in *ISPRS Annals* (II-5/W3), ed. by Yen, Y.-N., Weng, K.-H. and Cheng, H.-M., pp. 287–293.

24 Will, E. (1949a) La tour funéraire de Palmyre, in *Syria*, Vol. 26, pp. 87–116; see Silver, M., Törmä, M., Silver, K., Okkonen, J. and Nuñez, M. (2015) The Possible Use of Ancient Tower Tombs as Watchtowers in Syro-Mesopotamia, in *ISPRS Annals* (II-5/W3), ed. by Yen, Y.-N., Weng, K.-H. and Cheng, H.-M., pp. 287–293

25 See, for example, Kloner, A. and Zissu, B. (2007) *The Necropolis of Jerusalem in the Second Temple Period*. Interdisciplinary Studies in Ancient Culture and Religion 8. Leuven: Peeters.

26 The tower of Mariamne and Hippicus in the Citadel of Jerusalem; see Will, E. (1949b) La Tour Funeraire de la Syrie et les Monuments Apparentés, in *Syria*, Vol. 26, pp. 258–312.

27 Michalowski, K. (1964) *Palmyre, Fouilles Polonaises 1962*, Warszawa: Éditions scientifiques de Pologne, pp. 147–158.

28 See theories in Will, E. (1949b) La Tour Funeraire de la Syrie et les Monuments Apparentés, in *Syria*, Vol. 26, pp. 258–312, especially p. 302.

29 Amy, R. (1950) Temples à Escaliers, in *Syria*, Vol. XXVII, pp. 82–136.

30 See bibliography, for example, in Henning, A. (2010) The tower tombs of Palmyra: chronology, architecture and decoration, in *Studia Palmyreńskie*, Vol. 12, pp. 159–176.

31 See Lönnqvist, M., Törmä, M., Lönnqvist, K. and Nuñez, M. (2011) *Jebel Bishri in Focus: Remote sensing, Archaeological surveying, mapping and GIS studies of Jebel Bishri in central Syria by the Finnish project SYGIS*. BAR International Series 2230. Oxford: Archaeopress.

32 Will, E. (1949a)  La tour funéraire de Palmyre , in *Syria*, Vol. 26, pp. 87–116.

33 Will, E. (1949a)  La tour funéraire de Palmyre, in *Syria*, Vol. 26, pp. 87–116.

34 Henning, A. (2010) The tower tombs of Palmyra: chronology, architecture and decoration, in *Studia Palmyreńskie*, Vol. 12, pp. 159–176.

35 Henning, A. (2010) The tower tombs of Palmyra: chronology, architecture and decoration, in *Studia Palmyreńskie*, Vol. 12, pp. 159–176.

36 Will, E. (1949a) La tour funéraire de Palmyre, in *Syria*, Vol. 26, pp. 87–116.

37 Henning, A. (2010) The tower tombs of Palmyra: chronology, architecture and decoration, in *Studia Palmyreńskie*, Vol. 12, pp. 159–176, especially p. 160–161.

38 Watzinger, C. and Wulzinger, K. (1932)  Die Nekropolen, in Wiegand, T. (1932) *Palmyra –Ergebnisse der Expeditionen von 1902 und 1917*, Vol. 1, Text, Archäologisches Institut des deutschen Reiches, Abteilung Istanbul, Berlin: Verlag von Heinrich Keller, pp. 45–76, especially p. 48.

39 Wood, R. (1753) *The Ruins of Palmyra, Otherwise Tedmor in the Desart*. London: [Wood] .

40 Will, E. (1949a) La tour funéraire de Palmyre, in *Syria*, Vol. 26, pp. 87–116.

41 Brall, A., Breuer, M., Henning, A., Hohbuth, F., Prümm, O. and Stamm, T. (2001) Documentation of the Palmyrene Tower-Tombs in Syria Using Terrestrial Photogrammetry, in Archaeological and Photogrammetric Results, Proceedings of the XVIII International Symposium CIPA 2001, Surveying and Documentation of Historic Buildings – Monuments – Sites, Traditional and Modern Methods, Potsdam (Germany), September 18–21, 2001, Published by the CIPA 2001 Organising Committee, ed. by Albertz, J., *The International Archives of Photogrammetry, Remote Sensing and Spatial Information Sciences*, Vol. XXXIV, Part 5/C7, pp. 212–217.

42 See https://projectmosul.org. Accessed 20th November, 2016.

43 Will, E. (1949a) La tour funéraire de Palmyre , in *Syria*, Vol. 26, pp. 87–116; Silver, M., Törmä, M.,  Silver, K., Okkonen, J. and Nuñez, M. (2015) The Possible Use of Ancient Tower Tombs as Watchtowers in Syro-Mesopotamia, in *ISPRS Annals* (II-5/W3), ed. by Yen, Y-.N., Weng, K.-H. and Cheng, H.-M., pp. 287–293.

44 Schmidt-Colinet, A., Al-As'ad, Kh. and Al-As'ad, W. (2016) Palmyra, 30 Years of Syro-German/Austrian Archaeological Research (Homs), in *A History of Syria in One Hundred Sites*, Oxford: Archaeopress, pp. 339–348, especially  pp. 339–341.

45 Schmidt-Colinet, A., Al-As'ad, Kh. and Al-As'ad, W. (2016) Palmyra, 30 Years of Syro-German/Austrian Archaeological Research (Homs), in *A History of Syria in One Hundred Sites*, Oxford: Archaeopress, pp. 339–348, especially pp. 339–341.

46 Browning. I. (1979) *Palmyra*, London: Chatto & Windus, pp. 172–173.

47 Watzinger, C. and Wulzinger, K. (1932) Die Nekropolen, in Wiegand, T. (1932) *Palmyra– Ergebnisse der Expeditionen von 1902 und 191*, Vol. 1, Text, Archäologisches Institut des deutschen Reiches, Abteilung Istanbul, Berlin: Verlag von Heinrich Keller, pp. 62–69.

48 Watzinger, C. and Wulzinger, K. (1932) Die Nekropolen, in Wiegand, T. (1932)  *Palmyra – Ergebnisse der Expeditionen von 1902 und 1917*, Vols 1–2, Text und Tafeln,  Archäologisches Institut des deutschen Reiches, Abteilung Istanbul, Berlin: Verlag von Heinrich Keller,  pp. 71–76, pls. 38–44; Bounni, A. and Al-As'ad, Kh. (1997) *Palmyra: History, Monuments and Museum*, Damascus, p. 82; Browning, I. (1979) *Palmyra*, London: Chatto & Windus, pp. 176–179.

49 Bounni, A. and Al-As'ad, Kh. (1997) *Palmyra: History, Monuments and Museum*, Damascus, p. 82; Browning, I. (1979) *Palmyra*, London: Chatto & Windus, pp. 176–179.

50 Amy, R. and Seyrig, H. (1936) Recherches dans la nécropole de Palmyre, in *Syria*, Vol. 17, pp. 229–266.

51 Bounni, A. and Al-As'ad, Kh. (1997) *Palmyra: History, Monuments and Museum*, Damascus, pp. 98–100.

52 Cumont, F. (1909) *Les religions orientales dans le paganisme romain*. Conférences faites au collège de France en 1905. Annales du Musée Guimet. Bibliothèque de Vulgarisation Vol. XXIV. Paris.

53 Amy, R. and Seyrig, H. (1936) Recherches dans la nécropole de Palmyre, in *Syria*, Vol. 17, pp. 229–266.

54 Bounni, A. and Al-As'ad, Kh. (1997) *Palmyra: History, Monuments and Museum*, Damascus, pp. 98–100.

55 https://www.wmf.org/project/tomb-three-brothers. Accessed 29th September, 2016.

56 Sadurska, A. (1977) *Palmyre VII, Le tombeau de famille de 'Alainê*. Varsovie: Éditions scientifiques de Pologne.

57 Sadurska, A. (1977) *Palmyre VII, Le tombeau de famille de 'Alainê,*,Varsovie: Éditions scientifiques de Pologne, p. 74, see pp. 82–84.

58 Sadurska, A. (1977) *Palmyre VII. Le tombeau de famille de 'Alainê*, Varsovie: Éditions scientifiques de Pologne.

59 Rahmo, R. (1993) *Les Momies de Palmyre*, Damas: SIDAWY, p. 6.

60 Rahmo, R.(1993) *Les Momies de Palmyre*, Damas: SIDAWY, p. 24.

61 Grabar, O., Holod, R., Knudstad, J. and  Trousdale, W. (1978) *City in the Desert: Qasr al-Hayr East*. Vols. I, II. Harvard Middle Eastern Monographs 23-24. Cambridge, Mass.: Harvard University Press; *Chateaux Ommayades de Syrie, Collections du Musée National de Damas, 16 septembre 1990–17 mars 1991*. Paris: Institut du Monde Arabe; Lönnqvist, M., Törmä, M., Lönnqvist, K. and Nuñez, M. (2011) *Jebel Bishri in*

*Focus: remote sensing, archaeological surveying mapping and GIS studies of Jebel Bishri in central Syria by the Finnish project SYGIS*, BAR International Series 2230, Oxford: Archaeopress, pp. 343–349.

62 Rahmo, R. (1993) *Les Momies de Palmyre*, Damas: SIDAWY, p. 24–27.

63 Herodotos, *Histories*, Book II, 83–90.

64 Rahmo, R. (1993) *Les Momies de Palmyre*, Damas: SIDAWY, pp. 59–76.

65 Rahmo, R. (1993) *Les Momies de Palmyre*, Damas: S, Damas: SIDAWY, p. 27.

# 10

# THE CAMP OF DIOCLETIAN, THE CHRISTIAN BASILICAS AND THE ARAB CITADEL

## 10.1 The Camp of Diocletian and two temples

After the fall of Palmyra in AD 272, the city was transformed to accommodate a Roman garrison. During the tetrarchy of the Roman Empire, started by Diocletian in AD 293, new ideas were brought to Palmyra's planning. The western part of the city became increasingly occupied with public buildings and spaces in the 3rd century, although signs of occupation in the area with private houses date from the 1st and 2nd centuries AD. A Latin inscription commemorates the building of the camp during the reign of Diocletian, Constantine, and Maximian by Sosianos Hierocles, the governor of the province at the end of the 3rd century AD (CIL III, 133).[1]

As previously mentioned, the western part of the city came to be called the Camp of Diocletian in Late Antiquity (Figs. 10.1, 10.2).[2] The area was studied by early visitors, such as Robert Wood, who published drawings of ruined structures visible in 1753 (Fig. 10.3), and documented by D. Krencker in Theodor Wiegand's expedition in 1902 and 1917. The Polish excavations and studies have unearthed several structures that do not appear in early drawings.[3] The Diocletian quarters follow the plan of a Roman military camp, entered from the Grand Colonnade to a transversal colonnade called the Damascus Street and also approached via the *Porta Praetoria* to the northwest. The porticoed *Via Principalis* and the *Via Praetoria* crossed at another *Tetrapylon* at right angles.[4] Behind them was a Forum that

Fig. 10.1 *A panorama towards the Camp of the Diocletian taken from the south.*
Photo: Kenneth Lönnqvist 2004, SYGIS

195

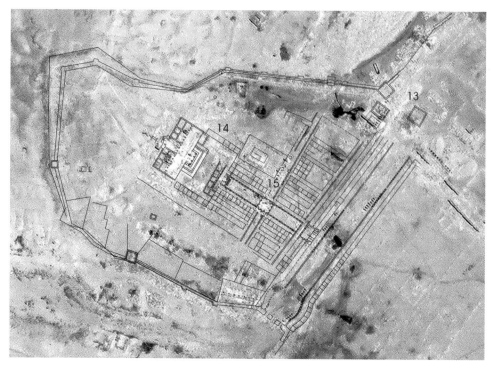

Fig. 10.2 *A satellite map of the Camp of Diocletian; note the barracks.* Courtesy of GORS. *No. 13. the Funerary Temple, no. 14. the Camp of Diocletian, no. 15. the Temple of Allat.*

Fig. 10.3 *The drawing illustrating the ruins of the Camp of Diocletian published by Wood in 1753.*

covered an area of 62 metres × 44.5 metres and served the Roman legionaries and administration. The area was entered by a Grand Gate with a portico during the 3rd century AD. Across the Forum one could approach the Temple of Signa (the Temple of Standards) and enter by a wide staircase and a portico.[5] This temple, that had served Roman legionaries, was drawn and published by Wiegand in 1932. The site was used by the Legio I Illyricorum that had participated in Emperor Aurelian's campaigns in the East. A headquarters known as the *Principia* and the *Praetorium* to accommodate the commander and military barracks have been unearthed in same the area[6] (see Figs. 10.2, 10.4).

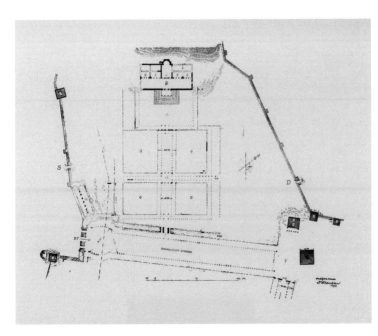

Fig. 10.4 *The Plan of the Camp of Diocletian showing the layout following a Roman military camp and the Temple of Signa published by Wiegand in 1932. Note the Temple of Allat was not excavated and visible yet, but the Temple of Signa is visible.*

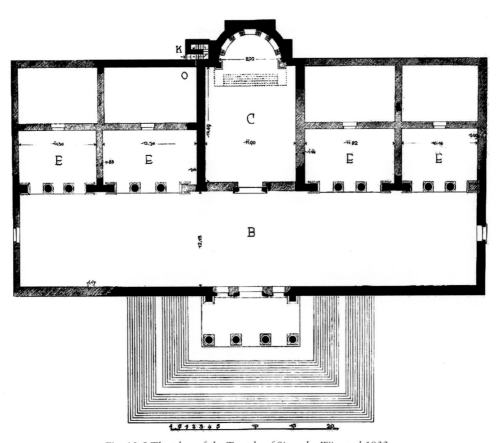

Fig. 10.5 *The plan of the Temple of Signa by Wiegand 1932.*

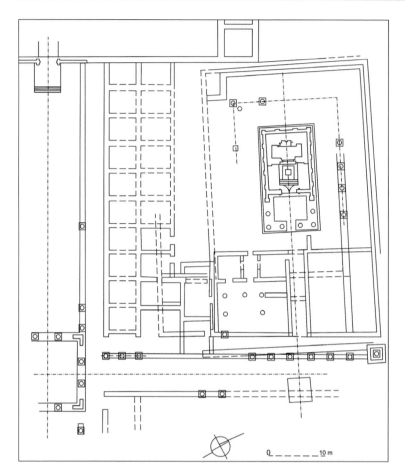

Fig. 10.6 *The plan of the Temple of Allat after the Polish Mission.* Redrawn by Ahmet Denker

Fig. 10.7 *The lion of Allat as it stood in the courtyard of the Palmyra Museum before ISIS.* Photo: Silvana Fangi 2010

Fig. 10.8 *The statue of Allat-Athena in the Palmyra Museum before the destruction by ISIS.* Photo: Kenneth Lönnqvist 2004, SYGIS

An earlier Temple of Allat flanked the Forum to the north (Figs. 10.2, 10.6). In the *temenos* of the temple there stood a large statue of a lion (Fig. 10.7) c. 3.5 metres in height and dating from the beginning of the common era.[7] It was moved to the entrance area of the Palmyra Museum. The statue was originally found in pieces and reconstructed. An Aramaic inscription associated with the lion promises Allat's blessings to all who do not shed blood in her temple.[8] After ISIS destroyed the lion in 2015[9] it was digitally modelled in 3D by Rekrei[10] and ICONEM.[11] Fortunately it hs been possible for the statue to be reconstructed by the Syrian heritage project under UNESCO.

The cult of the goddess Allat-Athena (for whom the other temple was dedicated) was a fusion of Oriental and Greek religions. Allat was an Arab goddess, who had a sister 'Uzza, from the Pre-Islamic times. Both were worshipped by nomadic Arabs and appear in the Nabatean Pantheon of Petra too.[12] Kazimierz Michalowski excavated the Temple of Allat in the 1960s, and later archaeological investigations were conducted by Michal Gawlikowski.[13] Behind the Sanctuary of Allat, in the Diocletian headquarters of the legions, an altar with an inscription and a relief representing a hand with a thunderbolt was discovered by Michalowski and dated to the time before the garrison, specifically AD 234.[14]

The Temple of Allat has corresponding elements with the Temple of Baalshamin. Both were erected outside the Grand Colonnade by the Bene Ma'azîn tribe and thus were connected with nomadic people. A bilingual inscription dedicated to Allat was erected at the site by the family in AD 64 (CIS 3966);[15] interestingly, this date matches up with the annexation of Syria to Rome. The Greco-Roman *cella* of the Temple of Allat has been dated by an inscription to the middle of the 2nd century AD, following the older sanctuary. The statue of goddess Allat imitating the famous cult statue of Athene by Phidias on the Acropolis of Athens originally furnished the temple and was in the Museum (Fig. 10.8). When Palmyra was recaptured it was found broken in the Museum in 2016 decapitated by the jihadists.

Over 70 sculptures including 60 portraits have been revealed from the ruins of the temple. There also were honorific sculptures on columns dating from the 1st century AD – before supporting brackets were used, sculptures had appeared in relief on the surface of a column. There is a column dedicated to Šalamallat, son of Yarhibol.[16]

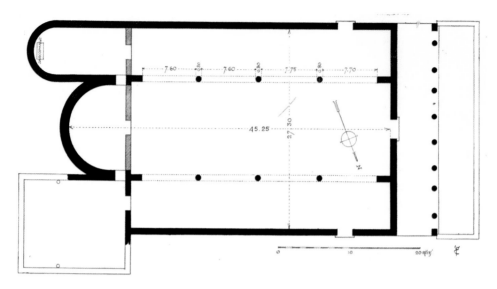

Figs. 10.9, 10.10 *Plans of two
Christian basilicas from Palmyra.
Note the portico in front of the
other.* Source: Gabriel 1926

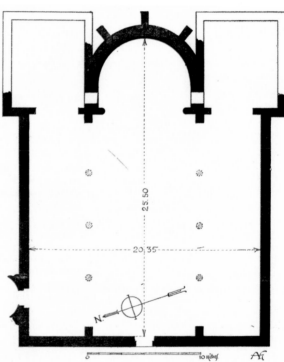

Fig. 10.11 *A stitched panorama to the ruins of Resafa-Sergiopolis.* Photo: Gabriele Fangi 2010

## 10.2 Christian basilicas

As mentioned, Christianity came early to Palmyra, as to Dura Europos. Several churches have been excavated or otherwise studied archaeologically in Palmyra.[17] We have seen how the Temple of Bel was also temporarily transformed in to a church (Ch. 6) like the Temple of Baalshamin.[18] Other churches in Palmyra follow the plan of a *basilica*. The *basilica* is a typical rectangular structure from the Roman period divided into a nave and aisles by columns, and as a church it has an apse in the cultic end. The construction of some churches in Palmyra involved reused and older architectural members.

One Christian basilica was identified by the German expedition led by Theodor Wiegand in 1902 and 1917, published in 1932.[19] Two Christian basilicas (Figs. 10.9 and 10.10) were also identified and documented by A. Gabriel at the beginning of the 20th century. They are situated north of the Grand Colonnade not far from the Temple of Baalshamin, on its western side. There also appears to be a reference in Gabriel's early maps to a synagogue in the neighbourhood. The larger Christian basilica was entered by a colonnaded porch. The plan follows Syrian basilicas and covers 27.30 metres in width and 45.25 metres in length. Six Corinthian monolithic columns divide the space into a nave and aisles. There probably was an iconostasis as well. The church does not date from earlier than the 4th century AD. The smaller church follows a similar plan covering 10.25 metres in width and 25.50 metres in length.[20] Both have rooms beside the apse that may have served a baptistery or a shrine for relics and a *narthex* or a porch.[21] Four churches have been excavated in Palmyra so far, and the identification of an episcopal complex has been made which includes the northernmost basilica, Basilica IV,[22] that had already been identified by the German expedition.[23] A bishop from Palmyra participated in the Council of Nicea in AD 325 and in the Council meeting of Chalcedon in AD 351.[24]

From the 5th century AD a petition to Emperor Leo mentions a bishop called Joannes from Palmyra.[25] During the reign of Emperor Justinian in the 6th century Christianity thrived in the city, and to that period also belong episcopal activities and pilgrimage in Resafa-Sergiopolis and Zenobia-Halabiya (see Ch. 5).[26] The huge enclosure of Resafa-Sergiopolis (Figs. 10.11 – 10.13), where the Saint Sergius' relics were venerated, situated

Fig. 10.12 *The large enclosure of Resafa-Sergiopolis.* Photo: Gabriele Fangi 2010

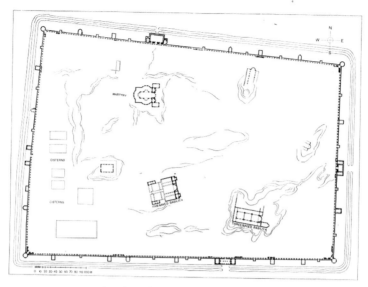

Fig. 10.13 *The plan of Resafa-Sergiopolis from 1928 before the modern excavations.* Source: Musil 1928

in the desert on the *Strata Diocletiana* while approaching the Euphrates. The area had been under the control of Palmyra, and Palmyra was one of the sites on the route. On the way of the pilgrimage several Christian tombs can be found in the desert; some were recorded and documented by the SYGIS project.[27] Ghassanids, Arab Christians, served Byzantium in the region during the 5th and 6th centuries AD, and their major pilgrimage site was Resafa-Sergiopolis. Ghassanids attributed the origins of Palmyra to King Solomon.[28]

## 10.3 Islam and the Arab castles

Islamic rule of Palmyra started with the Muslim conquests in the 7th century AD. Khaled ibn Al-Walid conquered Palmyra and its region in the 630s AD, probably AD 634. We have already mentioned the transformations of the Temple of Bel to a mosque and the fortification operation of its *Propylaea* during the Islamic period, and how a mummy was found in the city walls during the Umayyad period (AD 661–750). A mosque has also been discovered south from the *Tetrapylon* connected to the Grande Colonnade. It is a large building c. 42 metres × 26 metres, apparently dating from the Umayyad period.[29]

The mosque was partly built in the ruins of the Roman *peristyle* house or a small temple. While the Umayyads ruled the area, desert castles Qasr al-Hayr al-Gharbi (Fig. 10.14) and Qasr al-Hayr ash-Sharqi (Figs. 10.15, 10.16) were built in the region of Palmyra. That was the period of flourishing orchards in the desert utilising advanced water systems.[30] The Umayyad and Caliph Hisham's time had a particular art of its own in Syria-Palestine, with rich floral mosaics, such as in the Great Mosque in Damascus and Hisham's palace in Jericho in the Palestinian territories. During the Umayyad period a large *suq*, market place, with nearly 50 shops covered the area of the Grand Colonnade in Palmyra.[31]

Fig. 10.14 *The entrance to the National Museum in Damascus is in the form of a reconstructed gate to the Umayyad castle of Qasr al-Hayr al-Gharbi, the western castle.* Photo: Gabriele Fangi 2010

Fig. 10.15 *The Umayyad castle of Qasr al-Hayr ash-Sharqi, the eastern castle.*
Photo: Minna Lönnqvist 2000, SYGIS

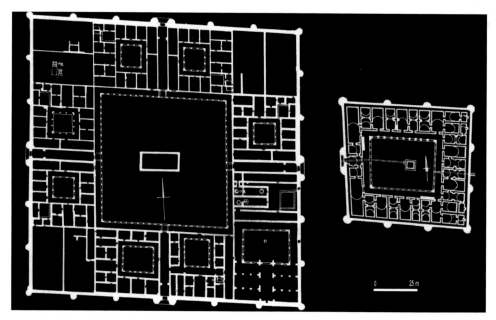

Fig. 10.16 *The plan of the Umayyad castle of Qasr al-Hayr ash-Sharqi after Grabar et al. 1978.*

(Above) Fig. 10.17 *Pottery from the Umayyad castles in the Palmyra Museum.* Photo: Kenneth Lönnqvist 2004, SYGIS. Courtesy of the Palmyra Museum

(Above right) Fig. 10.18 *The Arab castle of Palmyra photographed in 1920–1933.* Library of Congress, American Colony collection

(Right) Fig. 10.19 *The Arab castle of Palmyra photographed in 1920–1933.* Library of Congress, Matson Collection

The Umayyads were followed by the Abbasid (AD 750–1258) and Ayyubid (AD 1171–1260) dynasties. The Abbasid period took the focus away from the Palmyra region, but remains of Ayyubid structures can be found. During the Early Islamic period the Christians were included in the society and bishops were still chosen from the region.[32] Arab sources (Abu al-Farag) tell that Palmyra was destroyed during Caliph Abu al-Abbas in the 8th century AD. The Greco-Roman part of Palmyra was largely abandoned in the 9th century AD, and in the 10th century AD Palmyra was referred to as a *qasaba*.[33] There is a comparable site called Qasabi that had a caravan station in Medieval times on the Euphrates next to Halabiya.[34] If the reference to Palmyra as a *qasaba* reflects the form of an enclosure, that could well be applicable to its form inside the walls of Bel's Temple, where the Medieval village of Tadmor existed; we have seen the evolution of the temple courtyard filled with mudbrick houses (see Ch. 6). This is the view which Rabbi Benjamin Tudela apparently saw in the 12th century.

Now an Arab castle or citadel (Figs. 10.18–10.21) overlooks Palmyra on a rocky hill 150 metres above and north of the ruins from the Greco-Roman period. K. Wulzinger, a member of Theodor Wiegand's expedition, briefly described the fortress with drawings (published in 1932).[35] The castle is formed of several terraces and is surrounded by a moat. Fakhr ed-Din Ibn Ma'an (AD 1595–1634) was generally thought to have been the constructor of the castle, but the exploration and documentation of the castle in the 1990s brought new evidence to light. Some areas of the citadel have been dated to the Medieval period, apparently dating from the 12th and 14th centuries AD. The pottery includes famous Raqqa type of glazed sherds. Also the historical sources report that Fakhr ed-Din Ibn Ma'an conquered Palmyra in 1630 and strengthened the walls of the already existing castle. The western wing

Fig. 10.20 *The Arab castle or citadel before the damages caused by the civil war.* Photo: Silvana Fangi 2010

Fig. 10.21 *An entrance bridge to the Arab castle. The bridge has been destroyed during the civil war.* Photo: SYGIS 2000

Fig. 10.22 *Virtual model of the Arab castle of Palmyra with an imaginary landscape showing a wadi full of water like a river.* Constructed by Ahmet Denker

had been destroyed at some point by an earthquake or an enemy attack but was repaired during the Ottoman era. [36]

The castle has faced serious damage during the civil war, and ISIS took it for its use as a stronghold. Finally, the jihadists destroyed the bridge to the citadel. This citadel has also been digitally modeled in 3D by ICONEM[37], Arc/K project led by Brian Pope and by Ahmet Denker[38] (see Fig. 10.22).

## Endnotes

1  As'ad, Kh. and Yon, J.-B. (2001) *Inscriptions de Palmyre*, Promenades épigraphiques dans la ville antique de Palmyre, Guides archéologiques de l'IFAPO, No. 3, Beyrouth-Damas and Amman: Institut français d'archéologie du Proche-Orient, pp. 79–83.

2  See early studies by D. Krencker in Wiegand, T. (1932) *Palmyra – Ergebnisse der Expeditionen von 1902 und 1917*, Vol. 1, Text, Archäologisches Institut des deutschen Reiches, Abteilung Istanbul, Berlin: Verlag Heinrich Keller, pp. 85–105.

3  Krencker, D. (1932) Das Diocletianslager, in Wiegand, T. (1932) *Palmyra –Ergebnisse der Expeditionen von 1902 und 1917*, Vols 1–2, Text und Tafeln, Archäologisches Institut des deutschen Reiches, Abteilung Istanbul, Berlin, Verlag Heinrich Keller, see Vol. 1, Text: pp. 85–105.

4  Browning, I. (1979) *Palmyra*, Park Ridge, New Jersey: Noyes Press, p. 185.

5  Michalowski, K. (1961) *Palmyre: Fouilles Polonaises 1962*, Warsaw: Editions scietifiques polonaises, pp. 10–38; Michalowski, K. (1966) *Palmyre: Fouilles Polonaises 1963 and 1964*. Warsaw: Editions scientifiques polonaises.

6  As'ad, Kh. and Yon, J.-B. (2001) *Inscriptions de Palmyre*, Promenades épigraphiques dans la ville antique de Palmyre, Guides archéologiques de l'IFAPO No. 3, Beyrouth-Damas and Amman: Institut français d'archéologie du Proche-Orient, p. 82.

7  Bounni, A. and Al-As'ad, Kh. (1988) *Palmyra – History, Monuments and Museum*, Damascus, p. 60.

8  Drijvers, H.J.W. (1995) Inscriptions from Allât's sanctuary, in *ARAM*, Vol. 7, pp. 109–119, especially, p. 110.

9   https://www.theguardian.com/world/2015/jul/02/isis-militants-destroy-palmyra-stone-lion-al-lat. Accessed 28th September, 2016.

10  https://projectmosul.org/. Accessed 26th September, 2016.

11  https://sketchfab.com/models/02c4e194c6d64a4385a30990ed9899bf. Accessed 26th August, 2016; http://syri-anheritagerevival.org/museumpalmyra/. Accessed 25th November, 2016.

12  Christides, V. (2003) Religious Syncretism in the Near East: Allāt-Athena in Palmyra, in *Collectanea Christiana Orientalia* 1, pp. 107–116.

13  See the latest reports Gawlikowski, M. (2010) Palmyra, Preliminary report of the forty-fifth season of excavations, in *Polish Archaeology in the Mediterranean*, Vol. XIX, pp. 517–526.

14  Al As'ad, Kh. and Gawlikowski, M. (1997) *The Inscriptions in the Museum of Palmyra, A Catalogue*, Palmyra and Warsaw: Poland, p. 67.

15  As'ad, Kh. and Yon, J.-B. (2001) *Inscriptions de Palmyre*, Promenades épigraphiques dans la ville antique de Palmyre, Guides archéologiques de l'IFAPO, No. 3, Beyrouth-Damas and Amman: Institut français d'archéologie du Proche-Orient, pp. 80–81.

16  Drijvers, H.J.W. (1995) Inscriptions from Allāt's sanctuary, in *ARAM*, Vol. 7, pp. 109–119, especially, p. 110.

17  Wielgosz-Rondolino, D. (2016) Palmyrene portraits from the Temple of Allat, New evidence on artists and workshops, in *The World of Palmyra*, ed. by Kropp, A. and Raja, R., Palmyrenske Studier bind 1 - Palmyrene Studies Vol. 1, Scientica Danica, Series H, Humanistica, 4, Vol. 6, Viborg: Det Kongelige Danske Videnskabernes Selskab, pp. 166–179.

18  Gabriel, A. (1926) Recherches archéologiques a Palmyre, in *Syria*, Vol. VII, pp. 71–92, see especially Pl. XII and pp. 88–90; Gawlikowski, M. (2010) Palmyra, Preliminary report of the forty-fifth season of excavations, in *Polish Archaeology in the Mediterranean*, Vol. XIX, pp. 517–526; Genequand, D. (2008) An Early Islamic Mosque in Palmyra, in *Levant*, Vol. 40, pp. 3–15, especially p. 3 with the sources.

19  Browning, I. (1979) *Palmyra*, Park Ridge, New Jersey: Noyes Press, p. 169.

20  Wiegand, T. (1932) *Palmyra – Ergebnisse der Expeditionen von 1902 und 1917*, Vols 1–2, Text und Tafeln, Archäologisches Institut des deutschen Reiches, Abteilung Istanbul, Berlin: Verlag Heinrich Keller, see Vol. 2, Tafeln: Tafel 18.

21  Gabriel, A. (1926) Recherches archéologiques a Palmyre, in Syria, Vol. VII, pp. 71–92, see especially Pl. XII and pp. 88–90.

22  Browning, I. (1979) *Palmyra*, Park Ridge, New Jersey: Noyes Press, p. 170.

23  Gawlikowski, M. (2010) Palmyra, Preliminary report of the forty-fifth season of excavations, in *Polish Archaeology in the Mediterranean*, Vol. XIX, pp. 517–526.

24  Gawlikowski, M. (2010) Palmyra, Preliminary report of the forty-fifth season of excavations, in *Polish Archaeology in the Mediterranean*, Vol. XIX, pp. 517–526.

25  Genequand, D. (2008) An Early Islamic Mosque in Palmyra, in *Levant*, Vol. 40, pp. 3–15, especially p. 3 with the sources; see Constantine and the building of churches, for example, in Hunt, E.D. (1982) *Holy Land Pilgrimage in the Later Roman Empire A.D. 312–460*, Oxford: Clarendon Press.

26  Musil, A. (1928) *Palmyrena: A Topographical itinerary*. American Geographical Society. Oriental Explorations and Studies No 4. New York: American Geographical Society, p. 273; Musil, A. (1927) *The Middle Euphrates: A Topographical itinerary*, American Geographical Society, Oriental Explorations and Studies No 3, New York: American Geographical Society.

27  See Musil, A. (1928) *Palmyrena: A Topographical itinerary*, American Geographical Society, Oriental Explorations and Studies No 4, New York: American Geographical Society, pp. 299–326; Toivanen, H.-R. (2008) Fortresses and Ecclesiae on the Imperial Border of Byzantium, in *Jebel Bishri in Context, Introduction to the Archaeological Studies and the Neighbourhood of Jebel Bishri in Central Syria*, Proceedings of a Nordic Research Training Seminar in Syria, May 2004, ed. by Lönnqvist, M., BAR International Series 1817, Oxford: Archaeopress, pp. 111–126.

28  Lönnqvist, M., Törmä, M., Lönnqvist, K. and Nuñez, M. (2011) *Jebel Bishri in Focus: Remote sensing, Archaeological surveying, mapping and GIS studies of Jebel Bishri in central Syria by the Finnish project SYGIS*, BAR International Series 2230, pp. 255–261, 345.

29  Shahîd, I. (2009) *Byzantium and the Arabs in the 6th Century*, Vol. 2, part 2, Washington D.C.: Dumbarton Oaks Research Library and Collection, pp. 313–314; Lönnqvist, M., Törmä, M., Lönnqvist, K. and Nuñez, M. (2011) *Jebel Bishri in Focus: Remote sensing, Archaeological surveying, mapping and GIS studies of Jebel Bishri in central Syria by the Finnish project SYGIS*, BAR International Series 2230, p. 345.

30  Genequand, D. (2008) An Early Islamic Mosque in Palmyra, in *Levant*, Vol. 40, pp. 3–15, especially pp. 9–13.

31  Grabar, O., Holod, R., Knudstad, J. and Trousdale, W. (1978) *City in the Desert: Qasr al-Hayr East*. Vols. I,

II. Harvard Middle Eastern Monographs 23-24. Cambridge, Mass: Harvard University Press; *Chateaux Ommayades de Syrie, Collectons du Musée National de Damas, 16 septembre 1990–17 mars 1991*. Paris: Institut du Monde Arabe;  Lönnqvist, M., Törmä, M., Lönnqvist, K. and Nuñez, M. (2011) *Jebel Bishri in Focus: remote sensing, archaeological surveying mapping and GIS studies of Jebel Bishri in central Syria by the Finnish project SYGIS*. BAR International Series 2230. Oxford: Archaeopress, pp. 343–349.

32  Genequand, D. (2008) An Early Islamic Mosque in Palmyra, in *Levant*, Vol. 40, pp. 3–15, especially pp. 6–7.

33  Genequand, D. (2008) An Early Islamic Mosque in Palmyra, in *Levant*, Vol. 40, pp. 3–15, especially pp. 3–4, 13.

34  Genequand, D. (2008) An Early Islamic Mosque in Palmyra, in *Levant*, Vol. 40, pp. 3–15, especially pp. 3–4, 13.

35  See Qasabi in Lönnqvist, M., Törmä, M., Lönnqvist, K. and Nuñez, M. (2011) *Jebel Bishri in Focus: Remote sensing, Archaeological surveying, mapping and GIS studies of Jebel Bishri in central Syria by the Finnish project SYGIS*, BAR International Series 2230, pp. 215–217.

36  Wulzinger, K. (1932) Die islamische Burg bei Palmyra, Kal'at Ibn Ma'n, in Wiegand, T. (1932) *Palmyra – Ergebnisse der Expeditionen von 1902 und 1917*, Vols. 1–2, Text und Tafeln,  Archäologisches Institut des deutschen Reiches, Abteilung Istanbul, Berlin: Verlag Heinrich Keller, Vol. 1 Text: pp. 14–16, Vol. 2, Tafeln: Tafel: 8.

37  Bylínski, J. (1989–1990) The Arab Castle in Palmyra, in *Polish Archaeology in the Mediterranean*, Vol. II, pp. 91–93; see also http://www.maxvanberchem.org/en/scientific-activities/projects/?a=25 Accessed 31st August, 2016.

38  http://syrianheritagerevival.org/the-citadel-of-palmyra/. Accessed 25th November, 2016.

39  Scott, C. (2016) The Arc/k Project: Creating 3D Models to Preserve Threatened Landmarks for Future Generations - and for Hollywood, Aug. 18, in http://www.3dprint.com. Accessed 17th September, 2016.

# EPILOGUE

The lament for Palmyra is continuing in 2017. Caravans ended their travels and their trade to and from Palmyra ceased long ago. Palm trees are, however, still waving their leaves in the wind and the desert sand is blowing and whistling in the ruins, but the ancient ruins have been abandoned by tourists and modern life has not revived in the city. The historical city that was a vibrant centre of culture and tourism until 2011 has become more and more desolate, as invasions have followed time and again.

Since the manuscript for this book was completed in autumn 2016 the recapture of the city from ISIS, with the aid of the Russians, seemed to have halted the destruction of, and damage to, the ancient buildings. UNESCO, French and Polish researchers had opportunities to assess the state of the ruins during this interlude and portable items were taken to safety.

However, ISIS returned to Palmyra, reconquering it from the Syrian government forces in December 2016. The ancient city again became the target for destruction, and more damage to the ancient buildings was recorded. In January 2017 the Tetrapylon at the crossroads of the ancient streets was destroyed and structures at the Roman theatre were damaged, adding to those remains that had already vanished.

It was fortunate that no reconstruction or restoration work in the city had been started in haste during 2016 before the war has ended, because the situation obviously was not safe. The revival of the city in images, digital modelling and historical accounts has become a vital means of preserving what has been lost. Visual methods are aids to remember and they can help us to deal with a loss that is partly impossible to replace.

We shall continue to remember, wait and hope that Palmyra and its ruins will be reborn from the desert dust like a Phoenix from the ashes in the way that is appropriate to the conservation ethics and the past glory of the city.

Thank you for sharing with us this virtual journey though time to the ruins of Palmyra, by means of writing, images and architectural models. These carry and revive the memory of human life and cultural experience to the enrichment of us all.

# APPENDIX

## THE DOCUMENTATION OF ARCHITECTURAL HERITAGE IN SYRIA BY SPHERICAL PHOTOGRAMMETRY

### Gabriele Fangi

(edited by Minna Silver)

## The premise

In August 2010, the Crua (Recreational Club of the University of Ancona, now Polytechnic University of Marche) organized a tourist trip to Syria. I was in the group of tourists.

We visited the most important tourist and archaeological sites, such as the cities of Damascus, Aleppo, the dead city of Serjilla, the Roman city of Palmyra, Apamea, Shabba and Bosra, Crac de Chevaliers, and Qal'at Salah El-Din. At the time of the visit to Syria, I was just developing a photogrammetric technique, which I called spherical photogrammetry, characterized by the ease and speed of survey and simplicity of equipment.[1-4] Some details of the technique follow. With the newly developed technique, I had the chance to document 30 cultural items and emergencies in Syria (see the list below) just before the war. The surveys were partial and expeditious only and it could not have been otherwise then. The number of photographs obtained reached 17,000.

No one suspected that a fierce civil war would begin in 2011 and not withstanding the effect on the civilian population, the cultural heritage suffered total destruction and devastation. At the beginning of the war, I retrieved the photographic material collected and started to develop its metric contents. Only the results of the Palmyra projects are shown here.

While the photographic documentation has been done by myself, the restitution of Palmyra was performed by many of my students and in particular by Emanuele Ministri, Marco Franca, Chiara Manzotti, Clara Forino and Giada Francucci.

List of the sites visited:
Ancient City of Damascus
Ancient City of Aleppo
Ancient Villages of Northern Syria
Ancient City of Palmyra
Ancient City of Apamea
Ancient Cities of Shabba and Bosra
Crac des Chevaliers and Qal'at Salah El-Din

List of the documented emergencies:

Damascus – The Umayyad Mosque

Damascus – Al Sybahieh Mosque

Damascus – al Darwikiya Mosque

Damascus – Al Azem Palace

Damascus, National Museum – Qasr
al-Heir al-Gharbi gate

Damascus – The Sinan Mosque or
Tekkiye Mosque

Damascus – The minaret of al-Qali
mosque

Damascus – Damashen house

Damascus – Ottoman railway station

Aleppo – The Citadel walls

Aleppo – Khan el Wazir Gate

Aleppo – Bab al-Faraj Clock tower

Aleppo – Entrance gate of the
Ayyubid Palace in the Citadel

Aleppo – Al Helwieh madrasa inside
the citadel

Aleppo – Mihrab (madrasa
al-Halawiye)

Aleppo – San Simeon church

Aleppo – The minaret of the Umayyad
Mosque

Palmyra – Temple of Bel

Palmyra – Triumphal Arch

Palmyra – Roman Theatre

Palmyra –Tower Tomb of Elahbel

Serjilla, a dead city

Serjilla, a dead city

Apamea – the Roman temple to the
goddess Tyche

Shahba – Roman Theatre

Shahba – Filippeion

Bosra – Nabatean Arch

Bosra – Roman Theatre

Crac des Chevaliers

Hamah–minaret of al-Nuri

The projects related to Palmyra are synthesised in the following table:

| | Number of panoramas | No. of photos | Camera | Focal length | Resolution | Date | No. observations | Photographer | Restitution |
|---|---|---|---|---|---|---|---|---|---|
| Temple of Bel | 12 | 659 | Canon 450 D | 28 mm | 1504 × 2256 | 22/08/2010 | 336+175=511 | G. Fangi | G. Francucci |
| Triumphal arch | 3 | 50 | Canon 450 D | 28 mm | 1504 × 2256 | 22/08/2010 | 205 | G. Fangi | C. Forino |
| Theatre | 9 | 132 | Canon 450 D | 28 mm | 1504 × 2256 | 22/08/2010 | 349 | G. Fangi | E. Ministri |
| Elalhabel tower tomb | 12 | 132 | Canon 450 D | 50 mm | 1504 × 2256 | 22/08/2010 | 1948 | G. Fangi | M. Franca C. Manzotti |

The camera used was a Canon 450 D with 3 megapixel resolution, while the maximum resolution of the camera was 12 megapixel. The lens was a Canon 24–80 mm zoom. One could think that the equipment was not suitable for such a survey, but the low quality of the equipment compared with the excellent results are a good test and proof of the effectiveness of the technique, called here also 'Emergency Photogrammetry'.[5] Here I only provide an overview of the operations carried out and the results obtained, while the other authors give information about the archaeology.

## Spherical photogrammetry

Spherical photogrammetry was developed by Gabriele Fangi [2,3]. The name derives from the fact that the sensor is not the traditional flat plate (analogue or digital), but is a sphere, although a virtual one. From the same station point the operator takes a series of photographs, all with the same focal length, and partly overlapping, up to 360°. The photographs are downloaded to a computer, after which they are projected by means of popular commercial software over a virtual sphere. When the radius of the sphere and the focal length of the camera are equal, then the resolution on the sphere is equal to that of the original images, without loss of information. The sphere is then mapped on a plane with the equirectangular or azimuth-zenith or latitude-longitude cartographic projection. The image produced is called a spherical panorama. When the main axis of the sphere is vertical the two directions of the panorama are horizontal and vertical. When the $x$ axis is the representation of the equator of the sphere, the two image coordinates are longitude and latitude scaled by the radius of the sphere. In the digital images, the $x$ axis is the upper side, while the $y$ axis, facing downwards, is the left side. The representation equations are very simple

$x = \vartheta R$ and $y = \phi R$ with $R$ radius of the sphere and $\vartheta$ the horizontal direction with origin on the left side and $\phi$ the zenith direction.

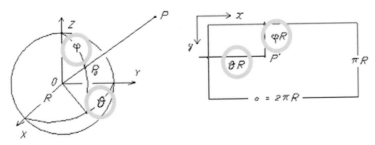

The flat image, i.e. the spherical panorama, is then the collection of all the possible outgoing directions from the station point.

Compared to a theodolite, an instrument capable of measuring the horizontal and the vertical directions, the difference is the lower resolution, however, compensated for by greater completeness. Furthermore, the sphere does not have any spirit level, and the axis of the sphere has a verticality error certainly greater than that of a theodolite. The verticality error needs to be corrected with two small rotations around the horizontal axes. For this reason, we speak of quasi-horizontal spherical panoramas.

Two adjacent views are linked together by means of the co-planarity equations. The set of the two projection centres and all the common collimated points form the so-called photogrammetric model. The series of formed photogrammetric models are connected together and referred to the same reference system by means of triangulation with independent models. This is a well-known and often used practice in photogrammetry, based on the similarity transformation equations. We need to supply a series of points with known coordinates, or control points, in the final reference system. The minimum requirements are one fixed point, known by its three coordinates, and three directions non-parallel to each other. However the independent model triangulation is not the final adjustment stage. The final adjusted coordinates and parameters are estimated in a bundle block adjustment based on the collinearity equations. A complete treatment of the procedure is given elsewhere.[6]

The advantages of this technique can be summarized as follows:

*Low-cost of the equipment.* In fact, the operator needs only a common digital camera, a tripod, a spherical head and a longimeter. Expensive instruments such as theodolites, total stations, laser scanning, GPS, or drones are not needed. Only a few distance measurements are needed. If reliable distance measurements are lacking, the model can be built with unknown size to be scaled later on, when suitable distance value will be available.

*Great speed of taking images.* No monumentation or targeting is required.

*Easy and safe transportation.* The equipment can be that of an ordinary tourist.

*Minimum bureaucracy and permissions.* In case of surveys with expensive instruments like total stations or laser scanning, the operator needs permission which is not easy to obtain in many countries. For instance in Syria it would have been impossible to carry out any survey.

*Possibility of using any topographic software.* The measures that can be derived from a spherical panorama are the same angular values that can be measured with a theodolite.

*Excellent photographic documentation.* The resolution of spherical panoramas is far superior to that of any camera available on the market.

*Field of view up to 360°.* All directions visible from a station are included in a spherical panorama.

*Possibility to have videos and animations.* The videos and the so-called quick-time are very common and valuable tools for cultural heritage documentation.

The disadvantages are summarised as followed:

*Full manual operations orientation and restitution.* At the moment all the collimating operations necessary for both the orientation of the panoramas and for the restitution are completely manual and entrusted to the experience and skill of the operator, making the plotting very slow.

*Difficult orientation of panorama.* In an age when the photogrammetric processes are tending toward 'one-click' programs, the orientation procedure is not trivial but requires sophisticated knowledge of photogrammetry and surveying.

*Decay of accuracy compared to traditional tacheometry.* A wide angle panorama normally reaches up to 30,000 pixels. Each pixel corresponds to an angle of $360/30,000 = 0.01°$ while a tacheometer has a resolution at least 10 times higher. Furthermore, the verticality error and the panoramas formation adversely affects the final accuracy. The spherical panorama is not 'set as a theodolite'.

*Absence of stereoscopy.* The collimation of the homologous points cannot occur for lack of stereoscopy, since the two panoramas forming a model do not allow stereoscopy. This greatly limits the number of points to be plotted to those easily identifiable in the different views. The procedure is suitable for the architecture while it is not for the plotting of natural objects or complex geometry.

For all these reasons and peculiarities, spherical photogrammetry has been defined here as 'emergency photogrammetry'. The final proof is the documentation of architectural heritage of Syria.

## The Temple of Bel

The surveys are composed of two different series of panoramas, one for the outside of the temple composed of 13 panoramic images and another for the interior of the temple composed of six panoramic images.

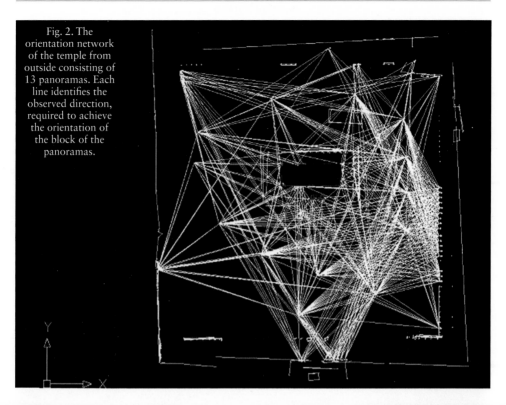

Fig. 2. The orientation network of the temple from outside consisting of 13 panoramas. Each line identifies the observed direction, required to achieve the orientation of the block of the panoramas.

3

4

5

6

7

Figs. 3–15. 13 panoramas, mostly 360° FOV (field of view).

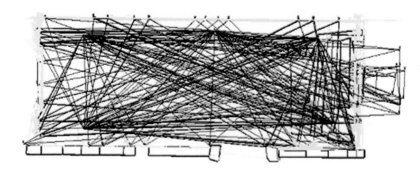

Fig. 16. The orientation network of six panoramas from the interior of the Temple of Bel.

17

18

19

Figs 17–22. The six panoramas from the interior of the temple.

**20**

**21**

**22**

Figs 23–28. Some views of the rendering of the temple of Bel by Giada Francucci.

## The Triumphal Arch of Palmyra

The survey was carried out by shooting five panoramas shown in the following figures (Figs 29–34). The fifth was not suitable for any plotting because it was the only one from its side. The arrangement of the other four is shown here.

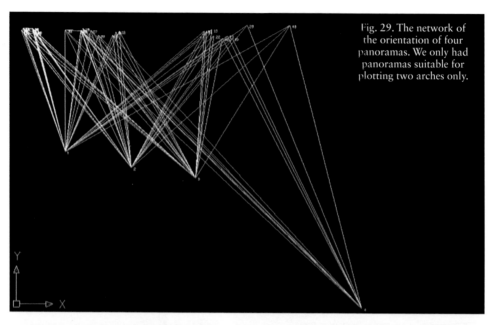

Fig. 29. The network of the orientation of four panoramas. We only had panoramas suitable for plotting two arches only.

Figs 30–34. The five panoramas that were prepared. The fifth is the
only one from its side of the arch, so did not not permit plotting.

33

34

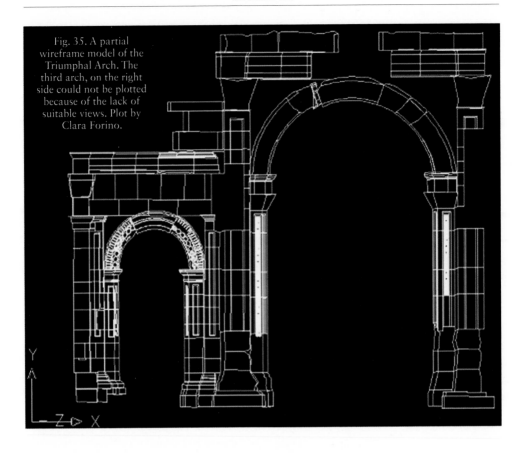

Fig. 35. A partial wireframe model of the Triumphal Arch. The third arch, on the right side could not be plotted because of the lack of suitable views. Plot by Clara Forino.

Figs 36 and 37. Solid models of the arch by Clara Forino.

Figs 38 and 39. Two views of the render of the arch by Clara Forino.

# The Theatre of Palmyra

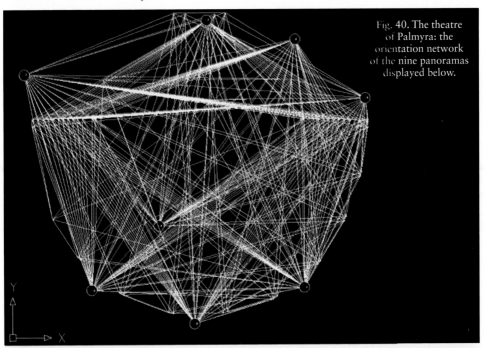

Fig. 40. The theatre of Palmyra: the orientation network of the nine panoramas displayed below.

Fig. 41. The theatre of Palmyra, an example of a spherical panorama composed of several photographs. The FOV (Field Of View) is not 360 °. The broken appearance of the sky is because it consists of several photographs stitched together.

42

Figs 42–49. The eight other panoramic images are shown here. The tourists are visiting the theatre.

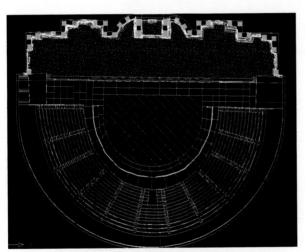

Fig. 50. The orientation plan of the theatre.

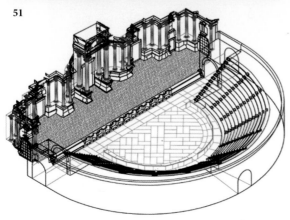

Figs 51–54. Views of the wireframe models of the theatre. Plot by Emanuele Ministri.

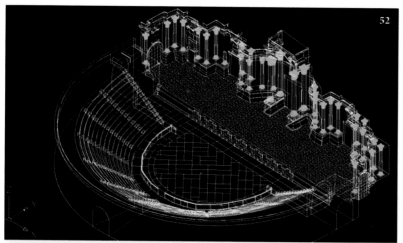

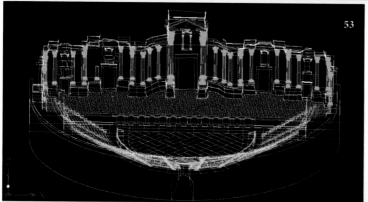

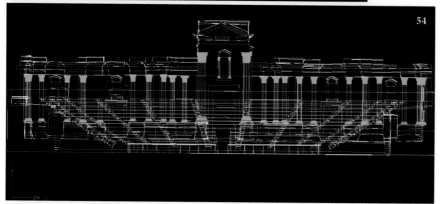

**56**

Figs 55 and 56. The Roman theatre of Shahba, a panorama and a wireframe model. This is a small theatre with only eight rows of seats. Plot by Guido Monachesi.

Figs 57 and 58. The Roman Theatre of Bosra. One of the five panoramas and a wire-frame plotting. Plot by Serena Freddo.

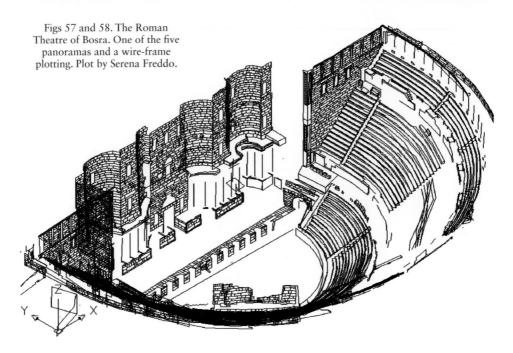

## The Elahbel tower tomb

The traverse around the tower is composed of 12 panoramas. None of them has a 360° FOV. The control network is shown in Fig. 59.

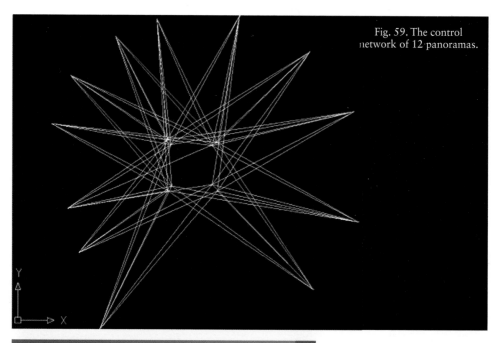

Fig. 59. The control network of 12 panoramas.

60

Figs 60–62. Three of the 12 panoramas of the tower tomb.

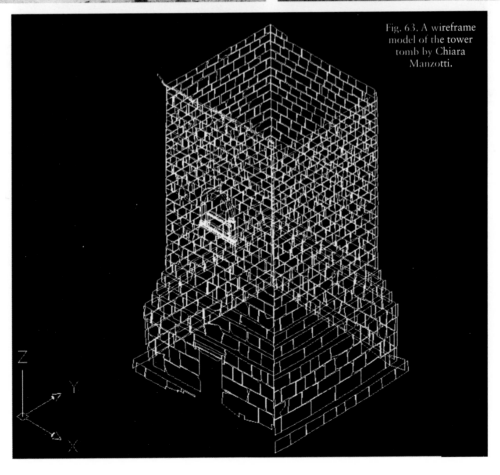

Fig. 63. A wireframe model of the tower tomb by Chiara Manzotti.

Figs. 64 and 65. Renderings of the tower tomb by Marco Franca.

# The formation of a database

All the projects have been stored in a database by means of dedicated cards in which all available information has been collected. We built a simple 'Object Table' the fields of which track some items such as: ID; Object name; Description (type, style, builder, date of construction, architectural analysis, intended use); Current state (type of damage); Localization (continent, nation, city, neighborhood); Different types of coordinate (latitude, longitude, WGS84, easting, northing UTM); Link to the products (panoramic images, documents, orientation, restitution, CAD, survey, multimedia) stored into a relative file path; Participants to the database project, customization, and other features.

# Acknowledgements

I deeply thank the students Giada Francucci, Clara Forino, Emanuele Ministri, Chiara Manzotti, Marco Franca, Francesco Di Stefano and the collegue Eva Savina Malinverni for their precious contribution and help.

# Further reading

Di Stefano, F. (2016) *Il patrimonio perduto della Siria, documentazione e archiviazione*. Graduation Thesis, Università Politecnica delle Marche, Anco.

Ministri, E. (2016) *Il teatro romano di Palmira con la fotogrammetria sferica: confronto con i teatri romani di Bosra e Sabratha*. Graduation Thesis, Università Politecnica delle Marche.

Fangi, G., and Pierdicca, R. (2012) Notre Dame du Haut by Spherical Photogrammetry integrated by Point Clouds generated by Multi-view Software, *International Journal of Cultural Heritage in the Digital Era*, Vol. I, n. 3, pp. 461–478.

Fangi, G., and Wahbeh, W. (2013) The Destroyed Minaret of the Umayyad Mosque of Aleppo, The Survey of the Original State, *European Scientific Journal*, Vol. 4, pp. 403–409.

Stierlin, H. (1987) *Cites du desert – L'art antique au Proche Orient*, Edité par Seuil, ISBN 10: 202009696X.

Wahbeh W., Neibiker S. and Fangi, G. (2016) Combining Public Domain and Professional Imagery for Accurate and Dense 3D Reconstruction of the Destroyed Bel Temple in Palmira, ISPRS Prague

# References

1    Fangi, G. (2007) The Multi-Image Spherical Panoramas as a Tool for Architectural Survey, 21th CIPA Symposium. In: *ISPRS International Archives*, Vol. XXX VI, Part 5/C53, *CIPA Archives*, Vol. XX I, pp. 311–316.

2    Fangi, G. (2009) Further Developments of the Spherical Photogrammetry for Cultural Heritage, 22nd CIPA Symposium, XX II CIPA Symposium, Kyoto, 11–15 October 2009.

3    Fangi, G. (2010) Multiscale Multiresolution Spherical Photogrammetry with Long Focal Lenses for Architectural Surveys, In: *ISPRS International Archives*, Vol. XXX VIII, Part 5, pp. 228–233.

4    Pisa, C., Zeppa, F. and Fangi, G. (2010) Spherical Photogrammetry for Cultural Heritage – San Galgano Abbey, Siena, Italy And Roman Theatre, Sabratha, Libya, In: *ACM Journal on Computing and Cultural Heritage*, Vol. 4, No. 3.

5    d'Annibale E., Piermattei L. and Fangi, G. (2011) Spherical Photogrammetry as Emergency Photogrammetry, – CIPA, Prague. *ISSN*: 1802-2669 ID 63761.

6    Fangi G. and Nardinocchi, C. (2013) Photogrammetric Processing of Spherical Panoramas, *The PhotogrammetricRecord* 28(143), pp. 293–311, DOI: 10.1111/phor.

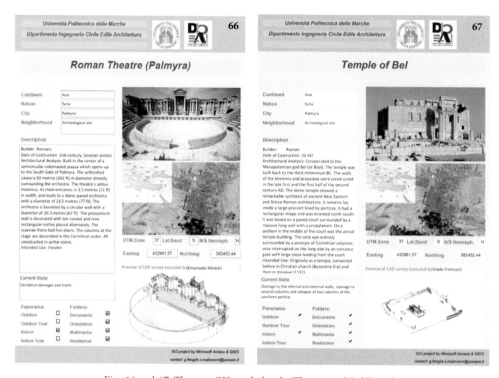

Figs 66 and 67. The two GIS cards for the Theatre and Bel Temple

# INDEX

(page numbers in italics refer to illustrations)

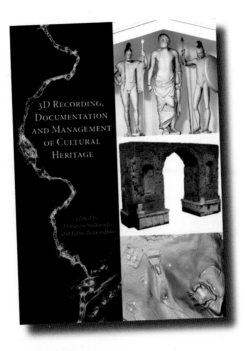